'O' Grade Biology

James Torrance

Edward Arnold

Preface

This book offers a complete collection of student
notes adhering closely to the O' Grade Biology
syllabus (but also taking into consideration the basic
needs of CSE and O' Level courses). It presents the
essential facts and ideas in the most concise and
visual manner possible. All areas of the O' Grade
course (including the recent syllabus changes) are
covered without the inclusion of irrelevant extras.
The book therefore performs several functions. It
acts as a useful tool of reinforcement and consolida-
tion throughout the class teaching of the entire
O' Grade course. This feature will be particularly
helpful to the average and the borderline pupil.
In addition it makes an excellent substitute for the
notes which many teachers issue, out of necessity,
to O' Grade pupils at present. Chapters are
deliberately short to allow for maximum flexibility in
teaching order. The book also comprises a compre-
hensive revision text of all the essential facts. In this
respect it is suitable for use by all O' Grade candidates
on completion of the course, especially since each
chapter ends with several revision questions in-
tended to reinforce and test the understanding of
fundamental ideas.

JT

© James Torrance, 1982

First published 1982
by Edward Arnold (publishers) Ltd
41 Bedford Square, London WC1B 3DQ

Reprinted 1982, 1983

British Library Cataloguing in Publication Data
Torrance, James
'O' Grade Biology
1. Biology
I. Title
574   QH308.7

ISBN 0 7131 0576 3

Diagrams by James Torrance

Typeset by Oxprint Ltd, Oxford
Printed in Great Britain

Contents

1 Variety and classification of living things

Biology is the study of living things. About two million different types of living organism are known to exist on the earth. In order to make the study of living things easier and to attempt to discover their basic origins and underlying relationships, biologists sort them into groups using a system of **natural classification**. Organisms are first placed in either the plant or animal kingdom and then the members of each kingdom are divided into several smaller groups.

Figure 1.1 Classification of the plant kingdom

(**Note**: Figure 1.3 shows the difference between radial and bilateral symmetry. A 'warm-blooded' animal is able to maintain constant body temperature regardless of its surroundings, whereas a 'cold-blooded' one is not.)

A locust belongs to the same enormous phylum (arthropods) as both a cockroach and a spider, since all three possess a hard exoskeleton and jointed legs, and to the same smaller class (insects) as a cockroach, since they both possess six legs. Possession of eight legs by the spider puts it into a different class (arachnids).

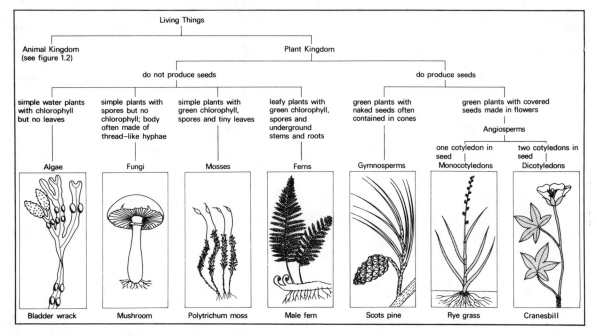

This information is often presented in a **branch key** as shown in figure 1.1 where the members of the plant kingdom are classified into six groups according to basic features such as production of spores (tiny, primitive reproductive units) or seeds (larger, more advanced reproductive units), presence or absence of green chlorophyll etc.

Similarly classification of the animal kingdom involves the use of fundamental structural characteristics (e.g. presence or absence of a backbone) rather than unreliable superficial features (e.g. ability to swim). Fairly similar members are grouped into large **phyla** with each phylum consisting in turn of several smaller **classes** whose members have many features in common and so on. Figure 1.2 shows a typical representative of each animal group and names the features used to classify it.

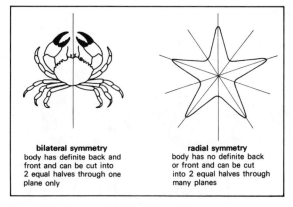

Figure 1.3 Types of symmetry

1

Figure 1.2 Classification of the animal kingdom

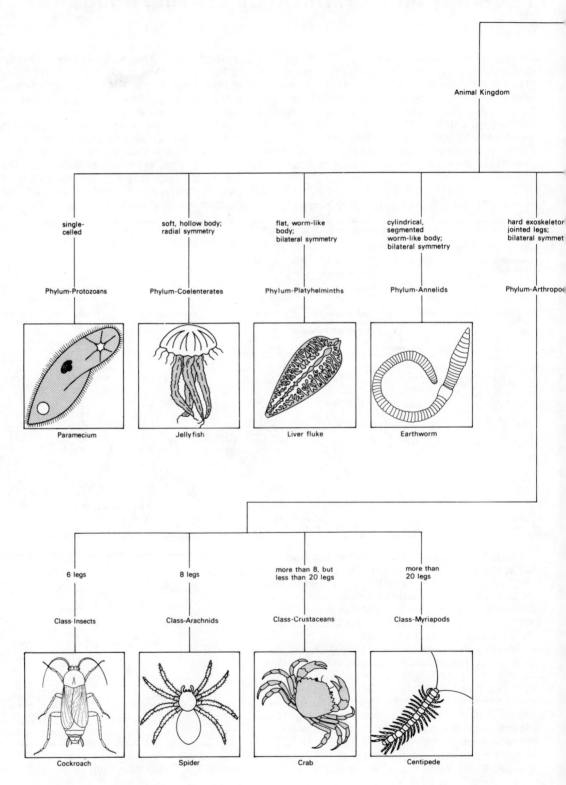

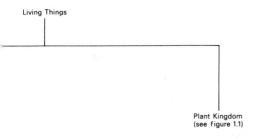

Living Things

Plant Kingdom
(see figure 1.1)

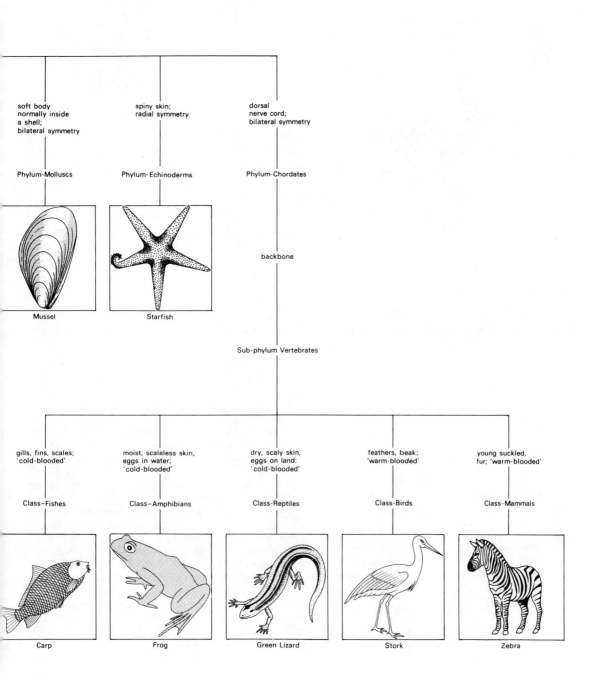

soft body
normally inside
a shell;
bilateral symmetry

spiny skin;
radial symmetry

dorsal
nerve cord;
bilateral symmetry

Phylum-Molluscs

Phylum-Echinoderms

Phylum-Chordates

backbone

Mussel

Starfish

Sub-phylum Vertebrates

gills, fins, scales;
'cold-blooded'

moist, scaleless skin,
eggs in water;
'cold-blooded'

dry, scaly skin,
eggs on land:
'cold-blooded'

feathers, beak;
'warm-blooded'

young suckled,
fur; 'warm-blooded'

Class–Fishes

Class–Amphibians

Class-Reptiles

Class-Birds

Class-Mammals

Carp

Frog

Green Lizard

Stork

Zebra

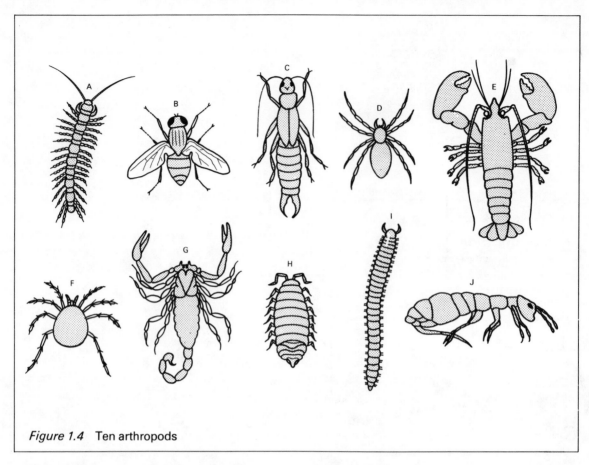

Figure 1.4 Ten arthropods

An alternative to a branch key is a key consisting of **numbered pairs of statements** as shown in table 1.1.

1	body bears 6 true legs	go to **2**
	body bears more than 6 legs	go to **4**
2	hind end bears 2 pincers	**C** (earwig)
	hind end bears no pincers	go to **3**
3	wings present	**B** (housefly)
	wings absent	**J** (springtail)
4	body bears 8 legs, head bears 2 pincers	go to **5**
	body bears more than 8 legs	go to **7**
5	body divided into 2 parts	go to **6**
	body not divided into 2 parts	**F** (mite)
6	sharp sting on end of abdomen	**G** (scorpion)
	no sting on end of abdomen	**D** (spider)
7	less than 20 legs present	go to **8**
	more than 20 legs present	go to **9**
8	10 legs present	**E** (lobster)
	14 legs present	**H** (woodlouse)
9	each segment bears 1 pair of legs	**A** (centipede)
	each segment bears 2 pairs of legs	**I** (millipede)

Table 1.1 Key to the ten arthropods shown in figure 1.4

Species and genus

A **species** is a small group of organisms which share so many common characteristics and are so similar to one another that they are able to interbreed and produce fertile offspring. Thus bulldogs, spaniels, dalmations, alsations and all other dogs belong to the same species.

Several different but closely related species such as dog, wolf, jackal and coyote which cannot interbreed to produce fertile offspring make up a larger group called a **genus** (plural genera). Similarly lion, tiger, leopard and jaguar are closely related species which belong to the one genus.

Every organism is named after both its genus and its species. Thus a dog's scientific name is *Canis familiaris*. This **binomial** (two name) system of classification was first introduced by a famous scientist called **Linnaeus** (1707–78) and several examples of it are given in table 1.2.

Genera belong to larger groups which in turn belong to even larger groups. Thus all the genera in table 1.2 belong to the same large group, the mammals, which in turn belongs to the even larger group, the phylum chordata.

common name	scientific name genus species
dog	Canis familiaris
wolf	Canis lupus
jackal	Canis aureus
coyote	Canis latrans
lion	Panthera leo
tiger	Panthera tigris
jaguar	Panthera onca
leopard	Panthera pardus
man	Homo sapiens

Table 1.2   Binomial system

Revision questions

1 (a) Which of the four animals shown in figure 1.5 is not bilaterally symmetrical?

(b) Pair each of the four animals with its closest relative in figure 1.2.

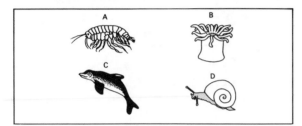

Figure 1.5   see question 1

2 (a) Green spleenwort is a spore-producing plant which has green leaves and underground stems. To which group of plants does it belong?

(b) In what way do fungi differ from all other plants?

3 Copy and complete the following table:

animal	phylum	class
millipede		
bat		
scorpion		
alligator		

4 When a horse is crossed with a donkey the result is a sterile animal called a mule. Do a mule's parents belong to the same species? Explain your answer.

5 Using the key given after this question identify the parent trees of the two winter twigs shown in figure 1.6 and then convert the key into a branch key.

(1)	smooth bark	go to (2)
	rough bark	go to (3)
(2)	buds cigar-shaped	beech
	buds not cigar-shaped	rowan
(3)	buds opposite	go to (4)
	buds alternate	lime
(4)	buds green	sycamore
	buds black	ash

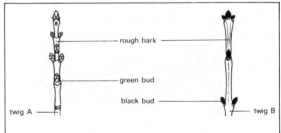

Figure 1.6   see question 5

2 Cells, tissues and organs

Every living thing is made of one or more **cells**. Typical animal and plant cells (figure 2.1) possess many of the same basic structures.

The **nucleus** (often seen more easily when stained with iodine solution) controls the cell's activities.

The **cytoplasm**, transparent jelly-like material, is the site of the chemical reactions essential for life. (Nucleus + cytoplasm = **protoplasm**).

The thin, flexible **cell membrane** enclosing the cell contents, controls which substances may enter and leave the cell.

The **cell wall** which bounds plant cells only, is made of **cellulose**. Although it is able to stretch slightly when water is absorbed by the cell, it is fairly rigid and therefore gives the plant cell definite shape.

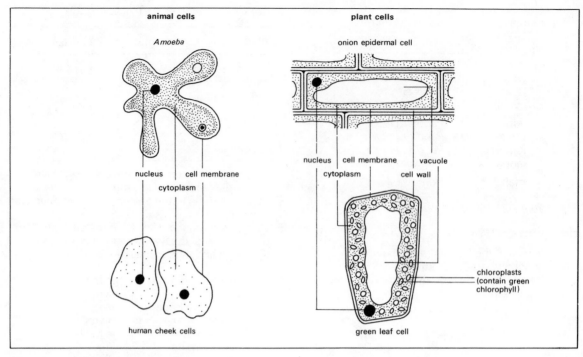

Figure 2.1 Typical animal and plant cells

A large permanent **vacuole** only occurs in plant cells. It often occupies most of the cell and always contains **cell sap** (a solution of sugars and salts). Some animal cells do however possess small temporary vacuoles (see figure 15.6).

Chloroplasts containing green **chlorophyll** are found in some plant cells where they carry out photosynthesis (see chapter 12).

Size of cells

When a small piece of graph paper bearing several pinholes, 1 mm apart, is viewed under the low power lens of a microscope, the pinholes are easily spotted since light passes through them. Thus the diameter of the microscope's field of view can be estimated; 2 mm for the microscope used in figure 2.2(a), for example. Next a specimen of cells is viewed and the average number of cells lying along the diameter of the field is found; 10 cells, for example, in figure 2.2(b).

Since **1 millimetre (mm) = 1000 micrometres (µm)**, the length of these 10 rhubarb epidermal cells = 2000 µm, and the average length of 1 rhubarb epidermal cell = 200 µm.

Plant cells are normally larger than animal cells. An onion epidermal cell, for example, is approx. 150 µm in length, a cheek cell is approx. 50 µm in length and a red blood cell (figure 2.3) about 7 µm in diameter.

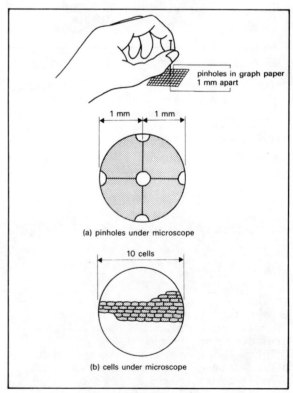

Figure 2.2 Measurement of cell size

Levels of organisation

Since a cell is the smallest unit that can lead
an independent life (an isolated part of a cell e.g.
nucleus does not survive), the cell is often described
as the basic unit of life. The unicellular animal
Amoeba neatly illustrates this idea by showing all
the characteristics of living things despite the fact
that it consists of only one cell.

In multicellular organisms, a **division of labour**
occurs amongst the cells. This means that instead of
every cell performing every function, certain cells
(figure 2.3) become specialised to perform particular
functions, thereby increasing overall efficiency.

A group of similar cells working together and
carrying out a specific function is known as a **tissue**
(e.g. muscle and bone). A group of different tissues
make up an **organ** (e.g. liver, eye and brain). A group
of related tisses and organs (e.g. heart, arteries and
veins containing blood) make up a **system** (e.g. circulatory system).

Integration within an organism

Rat
The thoracic cavity of the rat shown in figure 2.4
contains organs such as lungs and windpipe (parts
of the respiratory system) and heart and blood
vessels (parts of the circulatory system). The thorax

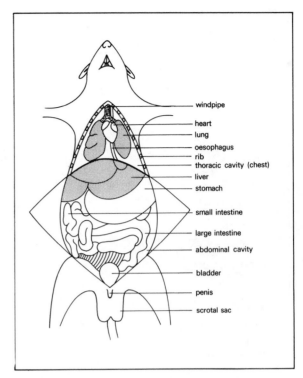

Figure 2.4   Dissection of male rat

is protected by ribs made of bony tissue. The
abdominal cavity contains the stomach and intestines
(organs belonging to the digestive system), the
kidneys and bladder (excretory system) and, in
females, ovaries, oviducts and uterus (reproductive
system). The male testes are found outside the
cavity in the scrotal sac. The entire body is supported
by the bony skeleton and co-ordinated by the
nervous system.

Flowering plant
Various levels of organisation also exist in plants.
The angiosperm (figure 2.5) has cells grouped into
tissues which, in turn, form organs such as root,
stem, leaf and flower. The root's functions are water
absorption and anchorage. The stem transports
food and water and supports the plant. The leaf is
responsible for photosynthesis and gaseous exchange with the surrounding air. The flower's function is reproduction and the seeds that it produces
eventually grow into the next generation of plants.

An advanced multicellular organism consists of
many different cells, tissues, organs and systems
each of which performs its own specialised func-
tion(s). For such a complex organism to lead an
independent healthy life, all of these systems must
operate in close co-ordination with each playing its
particular role as part of the **integrated whole**.

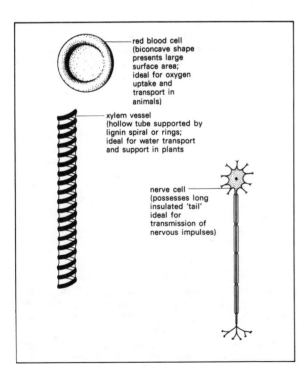

Figure 2.3   Specialised cells

7

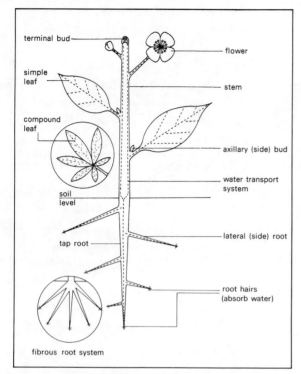

Figure 2.5   Structure of a flowering plant

Labels on figure:
terminal bud
simple leaf
compound leaf
soil level
tap root
fibrous root system
flower
stem
axillary (side) bud
water transport system
lateral (side) root
root hairs (absorb water)

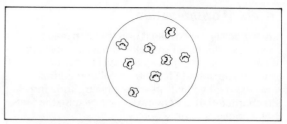

Figure 2.6   see question 2

2 Figure 2.6 shows white blood cells viewed under a microscope whose field of view is 150 μm in diameter. Estimate the diameter of a white blood cell.

3 Copy and complete the following table:

system	organs	function
digestive	stomach, intestines	
excretory		elimination of wastes
	lungs, windpipe	
		transport of blood
	ovaries, testes	production of sex cells
nervous		co-ordination

4 Arrange the terms **organ**, **system**, **cell**, **organism** and **tissue** in order of increasing complexity.

5 Explain why a red blood cell is often described as a good example of a cell whose structure is closely related to its function.

Revision questions

1 (a) Name 3 structures normally found in an animal cell and for each state its function.

  (b) Give 3 ways in which the leaf cell in figure 2.1 differs from *Amoeba.*

3 Composition of food

An **organic** food always contains the elements **carbon, oxygen** and **hydrogen**. Four classes of organic foodstuff are essential ingredients of man's diet.

Carbohydrates

These are energy-rich compounds often referred to as **'fuel'** foods. The simpler forms release their energy rapidly. Examples of foods rich in such **sugars** are fruit, honey and sugar cane. Complex carbohydrates are built up from repeating units of the simpler ones as shown in figure 3.1. Thus excess **glucose** (soluble) in plants is stored as **starch** (insoluble). Examples of foods rich in starch are wheat, rice and potatoes. Similarly excess glucose in animals (e.g. man) is stored as insoluble **glycogen** in the liver.

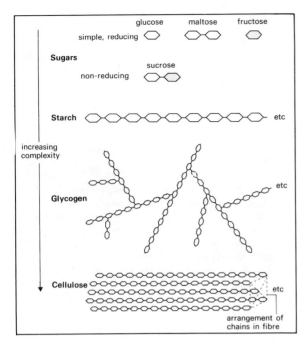

Figure 3.1   Types of carbohydrate

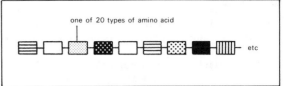

Figure 3.3   Small part of a protein molecule

The most complex carbohydrate shown in figure 3.1 is **cellulose**. It is made of thousands of glucose molecules arranged in long chains, five of which make up a tiny fibre. These, in turn, group together into larger cellulose fibres which make up the basic framework of plant cell walls. Thus all plant foods especially bran (husks of grain separated from flour after grinding) are rich in cellulose. Although man is unable to digest cellulose, it is an essential part of his diet because by giving bulk to his faeces, it stimulates the muscular action of the large intestine and therefore prevents constipation. Foods that have this effect are called **roughage**.

Fats and oils

Since **fat** (solid at room temperature) and **oil** (liquid at room temperature) release about twice as much

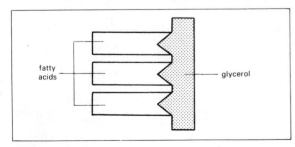

Figure 3.2   Fat molecule

energy per gram as carbohydrate, they are also **'fuel'** foods. Energy is released more slowly however from them than from carbohydrates. Fats are composed of two types of unit as shown in figure 3.2. Examples of foods rich in fat (or oil) derived from animals are butter, lard and cod liver oil; foods rich in fat (or oil) derived from plants are margarine, olive oil and palm oil. Excess fat is stored round man's kidneys and under the skin where it is especially effective in acting as a layer of insulation.

Proteins

In addition to the elements carbon, oxygen and hydrogen, proteins always contain **nitrogen**. Since excess protein cannot be stored by man, an adequate daily intake (about 80g) is required for body growth, formation of new cells and tissue repair. Examples of foods rich in protein are meat, eggs, fish, milk and soya bean. Though not normally a 'fuel' food, protein can release energy in a crisis (e.g. starvation). Each molecule of protein is made of many, many subunits called **amino acids** of which there are about twenty different types (figure 3.3).

vitamin	food rich in vitamin	deficiency disease or effect caused by lack of vitamin
A	carrot, milk, cod liver oil	poor night vision, dry skin
B$_1$	yeast, rice husks	beri-beri (paralysis of limbs)
C Ascorbic acid	green vegetables, citrus fruit	scurvy (soft gums, poor healing of wounds)
D	cream, cod liver oil	rickets (soft bones which become deformed easily)

Table 3.1   Vitamins

Vitamins

These organic foodstuffs (table 3.1) are only needed in tiny amounts. Some vitamins are destroyed by prolonged cooking.

Minerals

A wide variety of inorganic mineral salts are required by the human body. For example **calcium** (abundant in milk) is needed for healthy bones and teeth and **iron** (plentiful in lean red meat) is used to make haemoglobin, the red oxygen-carrying pigment found in red blood cells.

Food additives

Emulsifiers, preservatives, artificial sweetening, colouring and flavouring are just some of the 20 000 chemicals added to many of our foodstuffs. Although making the food last longer or seem more attractive, these substances rarely enhance the food's nutritional value and it is not yet known for certain if such food additives are definitely harmless when consumed regularly over a long period of time.

Balanced diet

A balanced diet is necessary for good health. This means both consuming an adequate amount of all the essential classes of foodstuff and taking in sufficient **water**. Water is an essential part of every cell's protoplasm and makes up about 70% of human body weight.

Food tests

Table 3.2 gives simple tests used to identify common foodstuffs.

Non-reducing sugars (e.g. sucrose) do not give a positive result with Benedict's solution. However if sucrose is boiled with dilute acid, this breaks it down to simple sugars which do react with Benedict's solution giving the orange colour.

The blue-black colour of dichlorophenolindophenol (DCPIP) is destroyed by vitamin C. If a solution is rich in vitamin C (e.g. fresh orange juice), then only a little of it is required to destroy the DCPIP's colour. If however the solution has a low vitamin C content (e.g. boiled orange juice), then a large amount is needed to destroy the DCPIP's dark colour.

Revision questions

1 Copy and complete the following table:

food	source	function
sugar		
roughage		
vitamin A		
iron		

2 Give 2 reasons why hibernating animals store fat rather than carbohydrate in their bodies.

3 Explain clearly why a pregnant woman needs to consume twice her normal minimum requirement of protein each day.

4 Consider the following results.

0.5 cm^3 of standard ascorbic acid solution decolourised 1 cm^3 DCPIP solution

1.0 cm^3 of tomato juice X were needed to decolourise 1 cm^3 DCPIP solution

10.0 cm^3 of tomato juice Y were needed to decolourise 1 cm^3 DCPIP solution

(a) Which tomato juice had been boiled? Give a reason for your answer.

(b) Calculate the ascorbic acid content of tomato juice X if the standard ascorbic acid solution contained 1 mg/cm^3.

5 Which (a) mineral (b) vitamin is required for healthy bone formation? (c) A diet deficient in these may lead to what deficiency disease developing?

6 Copy and complete the following table:

food	testing reagent	colour of positive result
glucose		
		blue-black
	DCPIP	

test	reagent (or material)	original colour of reagent	colour resulting when test is positive	heat or no heat	foodstuff present
1	Benedict's solution	blue	orange (brick-red)	heat	simple (reducing) sugar
2	iodine solution	brown	blue-black	no heat	starch
3	clean paper	white	greasy, translucent stain	no heat	fat or oil
4	Biuret reagent	blue	pale lilac	heat	protein
5	DCPIP* solution	blue-black	clear	no heat	vitamin C

Table 3.2 Food tests
(*DCPIP = dichlorophenolindophenol)

4 Energy content of food

When air is pumped into the apparatus shown in figure 4.1, the lid blows off the can showing that food contains energy which is released on burning.

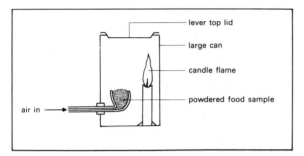

Figure 4.1   Exploding can experiment

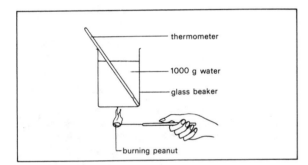

Figure 4.2   Measuring energy released

Measuring the amount of energy released from food

In the experiment shown in figure 4.2, 1 g of peanut was ignited and held under a beaker containing 1000 g water. The temperature of the water rose slightly (by about 2°C). Energy is measured in **kilojoules (kJ)**. **4.2 kJ** is the amount of energy required to raise the temperature of **1000 g** water by **1°C**. Thus 1 g of peanut has apparently released about 8.4 kJ. However much of the heat was lost to the surroundings, the heat that did reach the water was not evenly spread out and the peanut was not completely burned to ashes.

Food calorimeter
This apparatus (figure 4.3) is used to measure the energy content of a food and overcomes the difficulties described above. Since the food sample is

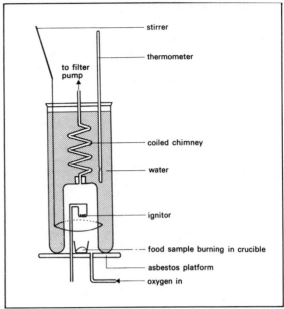

Figure 4.3   Food calorimeter

enclosed, little heat loss occurs. The **stirrer** and **coiled chimney** bring about even distribution of heat and the **oxygen supply** ensures that the peanut burns completely. This time the temperature of the water rises by about 6°C showing that 1 g of peanut releases about 25.2 kJ. The **calorimetric value** (amount of energy released when 1 g is burned) of various foods is given in table 4.1.

foodstuff	calorimetric value (kJ)
beans, baked	3.9
biscuits, sweet	23.5
bread, brown	10.1
bread, white	10.1
butter	33.5
cabbage	1.1
cheese, cheddar	17.8
cocoa powder	19.0
coffee powder	0.0
cornflakes	15.2
chicken	5.8
eggs, fresh	6.8
fish, white fried	8.4
fish, white steamed	2.9
kidney, fresh	4.2
margarine	33.5
meat, fresh beef	13.6
milk	2.8
peanuts	25.5
potatoes, boiled	3.7
potatoes, deep fried	10.0
tea, dry	0.0

Table 4.1   Calorimetric values

Energy requirements

When asleep, human beings need energy for keeping warm, for breathing, for heartbeat and for other workings of the body such as digestion and excretion. When awake, extra energy is needed for movement. The total amount of energy required per day depends on the **body size**, **age**, **sex** and **occupation** of the person (see table 4.2).

person	daily energy requirement (kJ)
2 year old child	5000
6 year old child	6500
12–15 year old girl	9600
12–15 year old boy	11700
woman (light work)	9500
woman (pregnant)	10000
woman (heavy work)	12500
man (light work)	11500
man (moderate work)	13000
man (very heavy work)	15500

Table 4.2   Daily energy requirements

When food in excess of the daily requirement is consumed, the body stores the extra as fat. To utilise this fat, and so lose the extra weight, the 'slimmer' must eat a diet in which the total energy content of the food is below the daily energy requirement.

Food shortage

Overweight tends to be a problem restricted to the wealthier peoples of the world. Two-thirds of the world's population barely receive sufficient food for their energy requirements. Even when food does contain enough energy, many people still suffer from **malnutrition** because their diet is unbalanced (i.e. one lacking in some essential food). Lack of vitamins, for example, leads to deficiency diseases. Protein malnutrition, common in many Third World countries, severely affects young children who should be growing quickly and often leads to the adult survivors becoming permanently weak and lethargic. Mental development in children may also be retarded since the brain does not develop properly under starvation conditions. Figure 4.4 shows how people in Central Africa often suffer from a shortage of protein (especially animal protein) whereas in Britain people eat much more than their daily protein requirement.

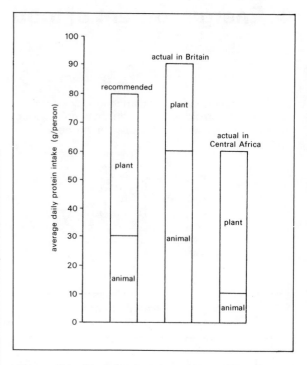

Figure 4.4   Comparison of protein intakes

Attempting to solve the problem

Although redistribution of foodstuffs seems an obvious answer, it normally only helps to solve the problem of world hunger under conditions of sudden famine caused by a disaster such as drought, flooding or crop failure. A long term solution requires a fairer distribution of the world's basic wealth. Developing countries need help to advance their **industry**, **commerce** and **agriculture**. They need stocks of disease-resistant plants and animals suitable for **breeding programmes**, supplies of useful **raw materials** and **education** for everyone so that the people can learn to help themselves. The basic problem of world hunger can only be tackled by changing the international economic system and by exercising birth control.

Revision questions

1 Since 1 g of peanut releases almost twice as many kilojoules as 1 g of glucose, of what class of organic foodstuffs must peanuts be mainly composed?
2 Sort out the following pairs of foodstuffs into those **(a)** suitable and **(b)** unsuitable for inclusion in a slimmer's diet: chicken and beef; cabbage and baked beans; boiled potatoes and chips; cocoa and black coffee.

3 Answer true or false to the following statements:
  (a) White bread is more fattening than brown bread.
  (b) Fried fish contains more energy than steamed fish.
  (c) Butter releases more kJ than margarine per gram.
  (d) 1 g of kidney would raise the temperature of 1000 g water by 1°C.
4 Match the following activities with the letters in figure 4.5:
  Sawing wood; sweeping; sleeping; building a brick wall.
5 By how many °C would 2 g of fried fish raise the temperature of 1000 g of water?

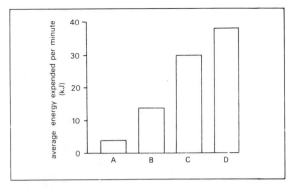

Figure 4.5   see question 4

5 Heterotrophic nutrition

Green plants which are able to convert light energy to chemical energy (stored in food) are called **autotrophs** (producers). Animals and non-green plants (e.g. fungi) which require a ready-made supply of chemical energy (stored in organic foodstuffs) are called **heterotrophs** (consumers).

Feeding in mammals

Mammals use teeth to prepare food for swallowing and digestion. Figure 5.1 shows the structure of a typical tooth. The numbers and types of teeth in a mammal are directly related to its diet.

Omnivore
An omnivore (e.g. man) eats both plant and animal material. To suit this mixed diet, the four types of teeth (figure 5.2) are all approximately the same size and lack the intense specialisation found amongst carnivores and herbivores.

Figure 5.2 shows how a **dental formula** presents concise information about exactly half of the total dentition. The dental formula of a child with a full set of milk teeth is i $\frac{2}{2}$ c $\frac{1}{1}$ p $\frac{2}{2}$ (total 20) since molars do not develop until the 32 permanent teeth grow.

Carnivore
A carnivore (e.g. dog) normally eats an exclusive diet of animal material (e.g. flesh) and its dentition (figure 5.3) is specialised accordingly. The long backward-curved canine teeth are especially suited

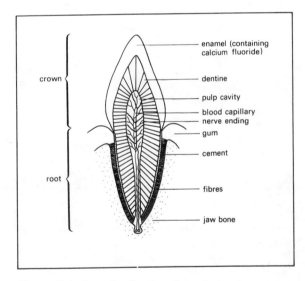

Figure 5.1   Longitudinal section of a tooth

to stabbing and holding prey. The premolars and molars bear cusps with sharp cutting edges. Since the lower jaw bites inside the upper jaw, a shearing action results on contraction of the powerful temporal muscle. The massive **carnassial** teeth gain leverage by being at the back of the mouth. They are such effective shears that they will even slice tendons and crack bones as well as cutting up flesh.

13

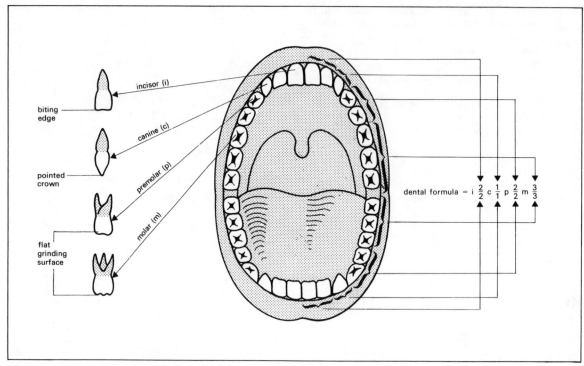

Figure 5.2 Dentition of adult human being (omnivore)

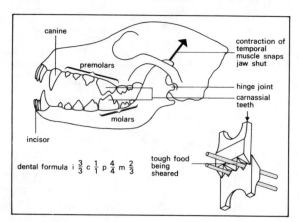

Figure 5.3 Dentition of dog (carnivore)

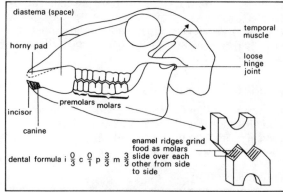

Figure 5.4 Dentition of sheep (herbivore)

Herbivore

A herbivore (e.g. sheep) eats an exclusive diet of plant material (e.g. grass) and its dentition (figure 5.4) is specialised accordingly. Incisor teeth in the lower jaw crop grass by biting against a horny pad on the upper jaw. In place of large canines, there is a space (diastema) where food is manipulated by the tongue. The loose nature of the hinge joint allows sideways movement of the lower jaw. This ensures that tough plant material is efficiently ground down by **enamel ridges** on the flat surfaces of the premolars and molars as they slide over one another.

Care of teeth

Bacteria in the mouth feed on sugar lodged between the teeth and produce **acid** which first dissolves the hard tooth enamel and then the dentine. The resulting hole enlarges until finally the soft pulp cavity is reached, nerve endings are exposed and pain is felt.

Teeth should be kept healthy by regular brushing to remove **plaque** (the deposit of bacteria, mucus and acid that gathers on the surface) by using a toothpaste which contains **alkali** to neutralise the harmful acid. Crisp foods (e.g. raw carrot) also remove food particles and calcium-rich foods (e.g. milk) help to build strong teeth.

14

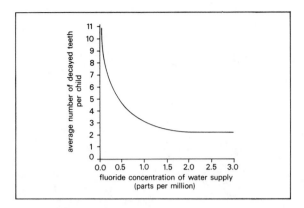

Figure 5.5 Effect of fluoride on tooth decay

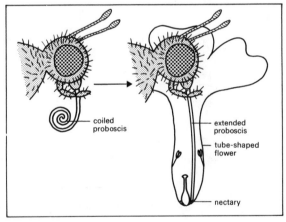

Figure 5.7 Feeding in housefly (sucking insect)

Fluoride

The graph in figure 5.5 shows the relationship between the amount of tooth decay and the concentration of fluoride in drinking water. Consideration of such data has persuaded many local authorities in Britain to raise the fluoride level of the local water supplies to 1 part per million in an attempt to reduce tooth decay. However this has become a controversial issue in many areas because some people object to being compelled to drink 'chemicals' which might, they claim, turn out to be poisonous over a long period.

Feeding in invertebrates

Biting insect

A locust cuts and grinds solid plant material using **mandibles** (figure 5.6). These main jaws, which are operated by powerful muscles, are assisted by smaller **maxillae**. Two pairs of **sensory palps** help to detect food.

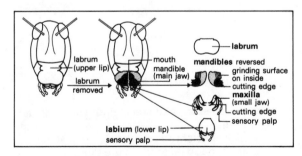

Figure 5.6 Mouthparts of locust (biting insect)

Sucking insects

A housefly's **proboscis** (figure 5.7) ends in two fleshy lobes channelled by tiny grooves. Saliva is secreted on to food which becomes dissolved and partially digested. This liquid food is then sucked up into the fly's gut. Flies feed on a variety of things including rotting garbage and animal faeces. If a contaminated fly lands on exposed food, germs from its proboscis, feet and faeces may be passed on to man causing disease.

A butterfly's mouthparts are shown in figure 5.8. During feeding this **coiled proboscis** is extended to suck up liquid nectar from flowers.

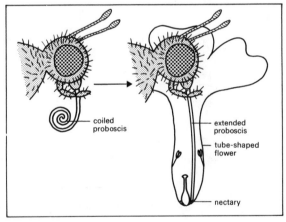

Figure 5.8 Feeding in butterfly (sucking insect)

Filter feeding

The beating of countless hair-like **cilia** on the **gills** of a bivalve mollusc (e.g. mussel) draws a current of water in through the **inhalent siphon** (figure 5.9). The water passes through pores (holes) in the gills, up vertical tubes and out by the **exhalent siphon**. Microscopic particles, however, become trapped in sticky mucus on the gill surface which acts as a filter. Those particles of nutritional value (e.g. algae) are passed to the mouth by ciliary action.

15

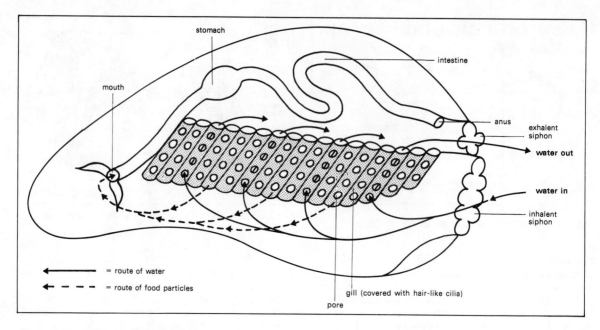

Figure 5.9 Filter feeding in mussel

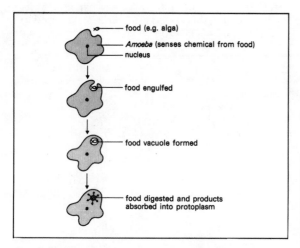

Figure 5.10 Phagocytosis in *Amoeba*

Phagocytosis
Amoeba feeds by engulfing and digesting smaller organisms as shown in figure 5.10. Further examples of heterotrophic nutrition are included in chapter 6.

Revision questions

1 In its upper jaw a horse has a total of 6 molars, 6 incisors, 6 premolars and 2 canines. The lower jaw contains the same number of teeth in the same ratio. Give the dental formula of a horse.
2 Describe a difference in structure and in function between the canine teeth of a dog and a sheep.

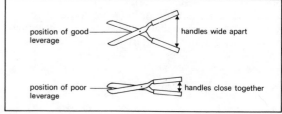

Figure 5.11 see question 3

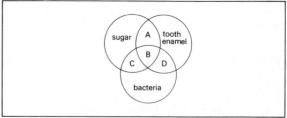

Figure 5.12 see question 4

3 Compare the shears shown in figure 5.11 with a dog's skull.
   (a) Which specialised teeth are located at the point of good leverage in the skull?
   (b) What does this enable these teeth to do very efficiently?
4 In which of the lettered regions in figure 5.12 will tooth decay occur?
5 Match the terms carnivore, sucking insect, filter feeder and herbivore with the examples oyster, zebra, mosquito and lion.

6 Symbiosis and saprophytism

A **symbiosis** is a close relationship where two different organisms live together. The organisms can be plants or animals. The three types of symbiosis are **mutualism**, **commensalism** and **parasitism**.

Mutualism

This is a symbiosis in which both of the organisms benefit.

Clover and Rhizobium
The **nodules** on the roots of plants with pods (leguminous) such as bean, pea and clover (figure 6.1) contain bacteria (*Rhizobium*) which fix atmospheric **nitrogen** into compounds used by the plant to make protein. In exchange, the bacteria receive carbohydrate from the plant.

Figure 6.3   Hermit crab and sea anemone

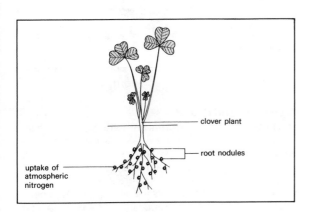

Figure 6.1   Root nodules

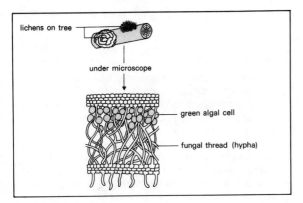

Figure 6.2   Lichens

Lichen
A lichen (figure 6.2) is a mixture of algal cells and fungal hyphae. The fungus receives sugar from the photosynthetic alga and the alga is kept moist by the water-retaining properties of the walls of the fungal threads.

Hermit crab and sea anemone
The anemone (figure 6.3) obtains free transport and scraps of food from the crab and the crab gains protection from the anemone's stinging cells.

Commensalism

This is a symbiosis in which one organism benefits and the other is neither benefited nor harmed.

Shark and remora fish
A remora fish (figure 6.4) has a sucker on its back and attaches itself to a shark in a position where it can pick up food scraps. The shark neither gains nor loses from the association.

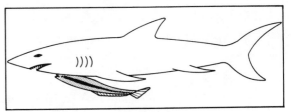

Figure 6.4   Shark and remora fish

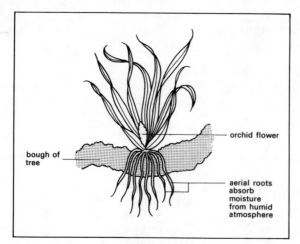

Figure 6.5 Tropical orchid

Tropical orchid and tree

The orchid (figure 6.5) is unable to survive at ground level in the tropical forest because it is too dark. It uses the branch of a tree as its base of attachmen⁺ and makes its food by photosynthesis. It neither harms nor benefits the tree.

Parasitism

This is a symbiosis in which one organism (the **parasite**) benefits at the expense of the other (the **host**).

Greenfly and plant

A greenfly (aphid) is an **external parasite**. Using its needle-like proboscis (figure 6.6) it pierces the plant tissues until it reaches the phloem from which it sucks out sugary juice thus depriving the plant of some of its food.

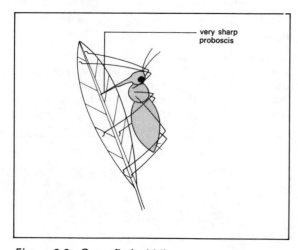

Figure 6.6 Greenfly (aphid)

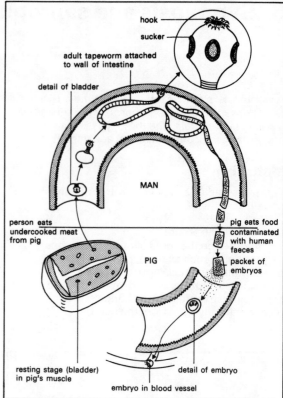

Figure 6.7 Life cycle of tapeworm

Tapeworm and mammal

The tapeworm is an **internal parasite**. Some, such as the pork tapeworm (figure 6.7), have more than one host. Such parasites are well adapted to suit their life style. They lack organs such as eyes and complex digestive systems since these are unnecessary, but do have suckers and hooks for clinging to the host.

Often a degree of **tolerance** exists between a host and a parasite. Although the host suffers (a person with a tapeworm becomes very thin and unhealthy) the well-adapted parasite will rarely do sufficient harm to kill the host.

Saprophytism

A saprophyte is an organism that obtains its food from dead or decaying organic matter. It secretes enzymes on to the food thus digesting it externally before absorbing the products into its body.

Many toadstools (figure 6.8), fungal moulds (e.g. *Mucor*, chapter 30) and bacteria are saprophytes. Under natural conditions in the soil, they digest and decompose the bodies of dead organisms and release **mineral salts** which return to the soil maintaining its fertility. They also play essential roles in the carbon and nitrogen cycles (see chapter 35).

Revision questions

1 In the following table + means benefit, − means harm and 0 means no significant effect. Identify each form of symbiosis.

form of symbiosis	species 1	species 2
A	+	−
B	+	0
C	+	+

fungal hyphae secrete enzymes on to dead leaves

Figure 6.8 Saprophytic toadstools

2 Classify the following examples of symbiosis into 2 named categories:
  (a) Certain bacteria in man's colon feed on unwanted food releasing vitamin B absorbed by man.
  (b) Female mosquitoes suck human blood (to nourish their eggs) and infect people with malaria.
  (c) The life cycle of the Chinese liver fluke involves 3 different hosts (man, snail and fish).
  (d) Egyptian plover birds clean leeches from between the teeth of crocodiles.
  (e) Dodder is a flowering plant which grows attached to stinging nettle plants from which it obtains all of its food.
  (f) Certain fungi on the roots of pine trees aid water absorption by the tree and receive carbohydrate from it.
3 Suggest 2 ways in which man could prevent the pork tapeworm (figure 6.7) from completing its life cycle.
4 Many parasites enjoy a delicate balance with their host where they harm it but do not kill it. Suggest why.
5 (a) Are saprophytic moulds autotrophic or heterotrophic plants?
  (b) Describe the important role played by such fungi in a deciduous woodland.

7 Enzymes

Need for digestion

Look at figure 7.1. When the water ('bloodstream') surrounding the visking tubing ('gut wall') in the experiment is tested at the start of the experiment, it is found to contain neither starch nor reducing sugar. After 30 minutes, however, it gives a positive result when tested with Benedict's solution, but a negative result with iodine solution. This shows that whereas the smaller sugar molecules have passed through the 'gut wall' into the 'bloodstream', the larger starch molecules have been unable to do so. Thus to be of use to the body, food molecules must be small.

Digestion is the conversion of large insoluble molecules into small soluble molecules and it is brought about by the action of **enzymes**.

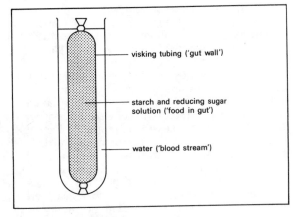

visking tubing ('gut wall')

starch and reducing sugar solution ('food in gut')

water ('blood stream')

Figure 7.1 Model gut experiment

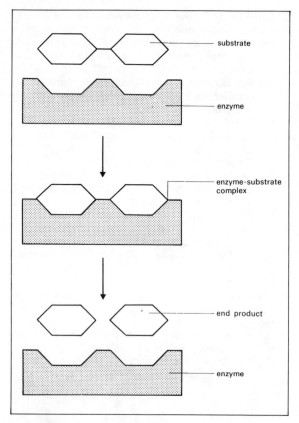

Figure 7.2 'Lock-and-key' theory of enzyme action

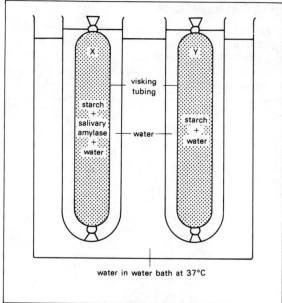

Figure 7.3 Action of salivary amylase

Enzymes

A catalyst is a chemical which speeds up the rate of a chemical reaction but is itself unchanged. An enzyme is a non-living protein made by living cells which speeds up the rate of a biochemical reaction but remains unchanged by the reaction. Enzymes are, therefore, **biological catalysts**.

An enzyme acts on only one type of substance (the **substrate**) and is said therefore to be **specific**. The **'lock-and-key'** theory of enzyme action is illustrated in figure 7.2. An enzyme is thought to have a shape that exactly matches its substrate allowing the two to combine briefly and the reaction to occur.

Every biochemical process in a living organism is enzyme-controlled. Thus many enzymes occur inside the cells whereas others promote reactions outside the cells (e.g. in the stomach). Some enzymes promote the synthesis (building up) of complex molecules from simpler ones (see figure 12.10); others control the breakdown of complex substances into simpler ones. The enzymes referred to in this chapter belong to the latter category since they digest food.

Action of salivary amylase (ptyalin) on starch

Salivary amylase is an enzyme contained in saliva. When the water surrounding visking tubing bag X (figure 7.3) is tested after one hour, it is found to contain reducing sugar, whereas that surrounding bag Y is still found to lack sugar.

Control
In this experiment bag Y is said to be the **control**. A control is a set of apparatus in which all of the conditions, except the one being investigated, are kept exactly the same as the original experiment. This allows the results to be compared. Any difference found between the two must be due to the condition under investigation. For example from the experiment in figure 7.3 we can conclude that starch (the substrate) has been digested by salivary amylase (the enzyme) to reducing sugar (the product). If a control had not been set up then it would be valid to suggest that perhaps starch would have become sugar anyway whether amylase had been present or not.

Effect of temperature on the activity of salivary amylase

At the start of the experiment shown in figure 7.4, no reducing sugar is found to be present in any of the test tubes when a sample from each is tested with Benedict's solution. After a few minutes each test tube is tested again. Although A and D still lack

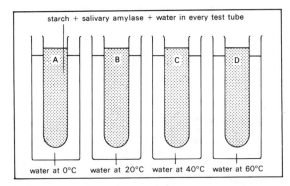

Figure 7.4 Effect of temperature on salivary amylase

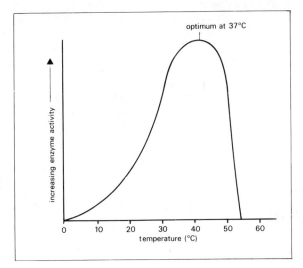

Figure 7.5 Effect of temperature on an enzyme

minutes the contents of tube A are clear whereas B and C remain unchanged. It is therefore concluded that pepsin digests large particles of insoluble protein to smaller particles of soluble semi-digested protein (peptides) only under acidic conditions.

The graph in figure 7.7 shows that for each enzyme there is a particular pH, the **optimum pH**, at which it works best.

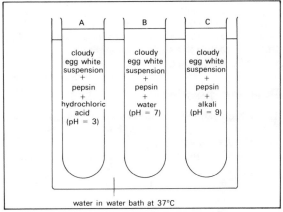

Figure 7.6 Effect of pH on pepsin

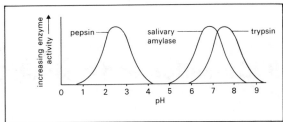

Figure 7.7 Effect of pH on enzymes

simple sugar, B is now found to contain a little (turns pale orange with Benedict's) and C has a lot of reducing sugar (bright orange). This experiment shows therefore that salivary amylase digests starch to simple reducing sugar most rapidly at around **body temperature**, slowly at room temperature and not at all at extremes of temperature.

This rule applies to enzymes in general as illustrated by the graph in figure 7.5. Enzymes are permanently destroyed (**denatured**) at temperatures over **50°C**.

Effect of pH on the activity of pepsin

Pepsin, an enzyme secreted by the stomach wall, digests protein. The protein used in the experiment shown in figure 7.6 is egg-white (albumen) which is a cloudy suspension of insoluble particles. After 30

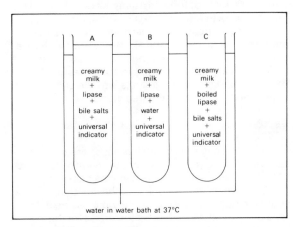

Figure 7.8 Action of lipase

21

Action of lipase on fat in creamy milk

The experiment in figure 7.8 is set up and then a little dilute alkali is added to each tube so that the contents begin at pH 8–9 (green). After an hour the contents of tube A have turned red (pH 3–4), tube B to yellow (pH 5) and tube C (the control) remains unchanged. It is concluded therefore that lipase, in the presence of bile, rapidly digests fat (or oil) to acids (fatty acids) whereas lipase without bile only slowly digests fats to fatty acids.

Bile is not an enzyme. It speeds up the reaction by converting large drops of fat into tiny globules of fat, thus increasing the surface area of substrate upon which lipase can act.

Revision questions

1 In which of the 4 test tubes shown in figure 7.9 will the cylinder of hard-boiled egg-white disintegrate fastest? Explain why.
2 (a) When tube A in figure 7.8 is repeated using pepsin instead of lipase, there is no reaction. What property of an enzyme does this illustrate?
   (b) State 2 other properties characteristic of enzymes.

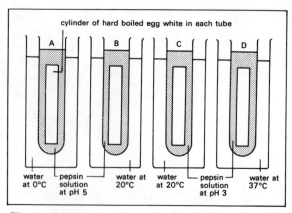

Figure 7.9   see question 1

3 Briefly explain each of the following:
   (a) A control experiment is often referred to as a 'standard of comparison'.
   (b) Dry maize grains contain starch but no reducing sugar whereas soaked maize grains that have been germinating for 3 days do contain reducing sugar.
   (c) Fevers which raise the body temperature to over 42°C are normally fatal to human beings.

8 Digestion and assimilation

Living cells throughout the human body need a constant supply of food (in a simple soluble state) for energy, growth and repair. Digestion, the breaking down of large insoluble molecules into smaller soluble ones, occurs in the **alimentary canal** (figure 8.1), a muscular tube running from mouth to anus.

The salivary glands, liver and pancreas all of which pass secretions via ducts into the alimentary canal are said to be associated **glands**.

Mouth

Chewing food in the mouth increases the **surface area** upon which enzymes (chapter 7) can act. Saliva made in the salivary glands contains **salivary amylase** which digests some of the starch to reducing sugar (maltose) as shown in figure 8.2a. On being swallowed, food is forced down the oesophagus by muscular activity called peristalsis

(figure 8.3). This process occurs throughout the entire gut by circular muscle in the gut wall relaxing and contracting alternately.

Stomach

Closure of the muscular ring (**sphincter**) at each end of the stomach holds food there while vigorous muscular activity of the stomach wall (figure 8.4) churns the food up with **gastric juice** secreted by the glands in the stomach wall. This juice contains the enzyme **pepsin** (which digests protein to peptides, figure 8.5a), **hydrochloric acid** (which kills germs and creates the low pH conditions needed by pepsin) and **mucus** (a slimy liquid formed in most regions of the gut which aids the movement of food and prevents the gut wall from being digested by its own enzymes).

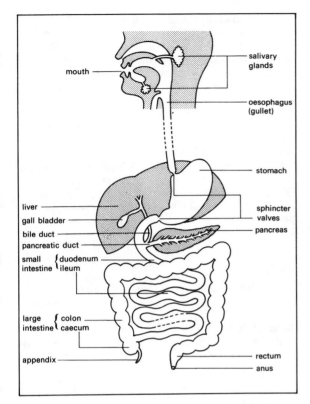

Figure 8.1 Human alimentary canal (gut)

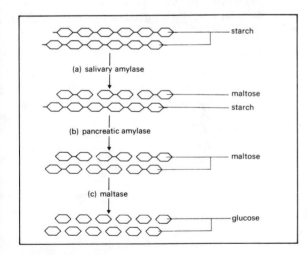

Figure 8.2 Carbohydrate digestion

Small intestine

As semi-digested food is squirted a little at a time into the duodenum, it meets green alkaline **bile** and **pancreatic juice**.

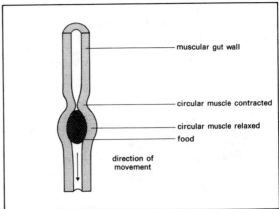

Figure 8.3 Peristalsis

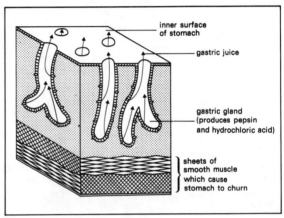

Figure 8.4 Structure of stomach wall

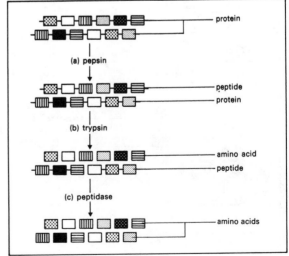

Figure 8.5 Protein digestion

23

Bile is made in the liver and does not contain enzymes. It is stored in the gall bladder and passes down the bile duct into the duodenum where it neutralises acid from the stomach (thus creating conditions suitable for the pancreatic enzymes to work) and **emulsifies** large fat globules to tiny droplets (thus increasing the surface area for digestion).

Pancreatic juice contains three enzymes. **Amylase** (figure 8.2b) digests any remaining starch to maltose; **trypsin** (figure 8.5b) digests protein and peptides to peptides and amino acids; **lipase** digests fats to the end products fatty acids and glycerol (figure 8.6).

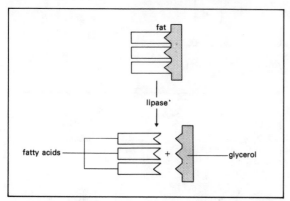

Figure 8.6 Fat digestion

The food undergoing this continuous digestion is forced by peristalsis into the ileum, the wall of which secretes further enzymes. These digest maltose to glucose (figure 8.2c) and peptides to amino acids (figure 8.5c).

Absorption of digested food in the ileum

The ileum (figure 8.7) is structurally suited to its role of absorption in several ways. Its long **tube-like** shape presents a large surface area to the soluble end products of digestion. This effect is further increased by the inner lining being folded and bearing many tiny finger-like **villi** (figure 8.7). In addition the layer of cells (**epithelium**) covering each villus is **thin** allowing food to diffuse through easily. Lastly each villus contains a rich blood supply and a tiny lymph vessel, the **lacteal**.

Fatty acids and glycerol pass into the lacteals and on into the lymphatic system (see chapter 16). Glucose and amino acids diffuse directly into the blood capillaries which unite to form the **hepatic portal vein** (figure 8.8). This vessel carries all blood leaving the gut to the liver.

Large intestine

Material passed into the large intestine consists largely of undigested matter (e.g. cellulose), bacteria and dead cells. The colon does not secrete enzymes. Its function is to absorb water from the unwanted material which becomes **faeces**. This waste material is stored temporarily in the rectum and later expelled through the anus.

Roles of the liver

1 **Formation of bile from old dead red corpuscles**
2 **Storage of iron and vitamins A and D**
3 **Breakdown of poisons e.g. alcohol**
4 **Regulation of blood sugar level**

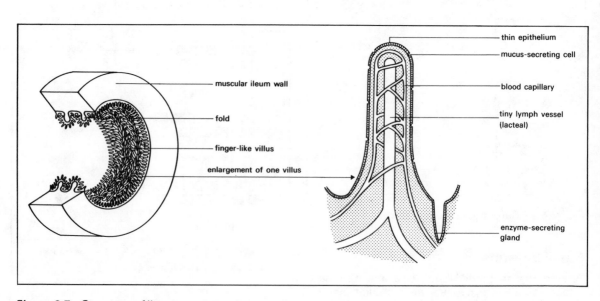

Figure 8.7 Structure of ileum

Excess glucose arriving by the hepatic portal vein (figure 8.8) is stored as glycogen. When the blood sugar level decreases, some glycogen is converted back to glucose to maintain the correct balance.

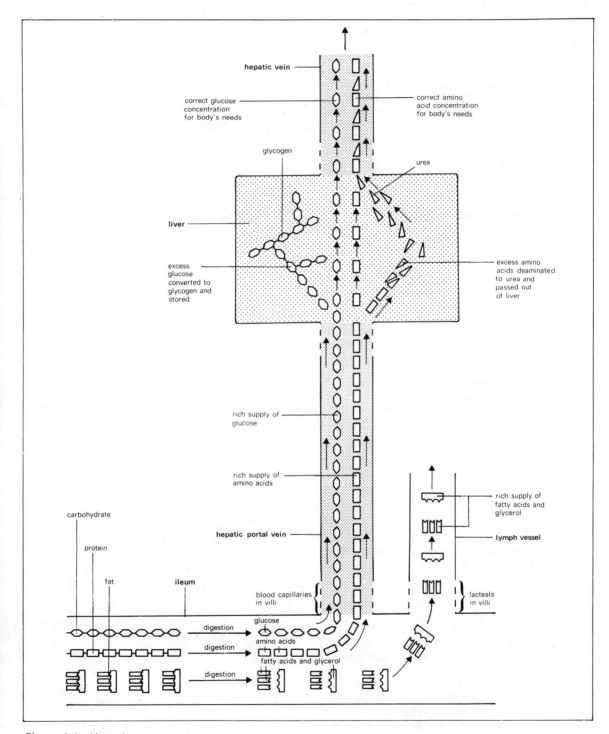

Figure 8.8   Hepatic transport

5 Deamination of surplus amino acids

Excess amino acids cannot be stored (they would become poisonous) and they are therefore broken down to the harmless waste product **urea**. This process is called **deamination**.

Blood entering the liver by the hepatic portal vein contains a lot of glucose and amino acids and a little urea. However, blood leaving the liver by the hepatic vein contains a medium concentration of glucose and amino acids exactly suited to the body's needs and a high concentration of urea. This blood passes into general circulation (see chapter 16). The urea is removed by the kidneys and the soluble foods are transported in blood plasma to every living cell where glucose is used for energy and amino acids are built up (**assimilated**) into the cell's proteins during growth and tissue repair.

Cellulose digestion in herbivores

A herbivore's diet consists largely of cellulose, the material that acts as roughage but remains un-digested in man's gut. Cellulose is digested to simple sugar in the gut of a herbivore; not by the herbivore's own enzymes but by the enzyme **cell-ulase** secreted by millions of **bacteria** living in the animal's gut. This is an example of mutualism since the cellulose-digesting bacteria (called the **intestinal flora**) receive warmth and food in return for their efforts.

In a cow these bacteria are found lining the **rumen**, the first of three 'stomachs' that precede the true stomach. In a rabbit these bacteria are found in a well developed region of the large intestine (**caecum**) which is small in humans (compare figures 8.1 and 8.9). Digestion of cellulose in the large intestine, where no absorption of soluble end products occurs, would seem of little value. However rabbits eat their faeces and therefore gain the sugars the second time around.

Revision questions

1 Which letter in figure 8.10 indicates the region in a pig's gut where:
   (a) fat digestion begins
   (b) many germs are destroyed
   (c) water is absorbed into the bloodstream
   (d) a sphincter valve is located
   (e) salivary amylase begins to play its role
   (f) the gut wall is covered with villi
2 Write a sentence to explain each of the following statements:
   (a) A small piece of bread begins to taste sweet if held in the mouth for 15 minutes.
   (b) Despite undergoing gastrectomy (removal of stomach), complete digestion of protein still occurs in man.

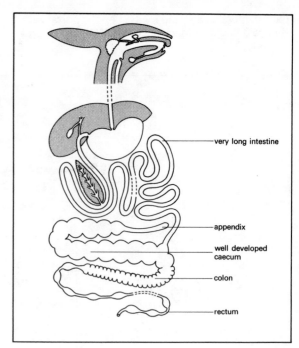

Figure 8.9   Rabbit's gut

- very long intestine
- appendix
- well developed caecum
- colon
- rectum

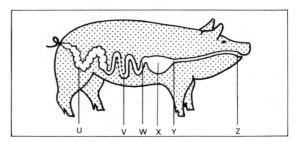

Figure 8.10   see question 1

   (c) Even after a meal rich in sugar, the hepatic vein still contains less glucose than the hepatic portal vein.
   (d) The walls of a cow's rumen contain no en-zyme-secreting glands yet food still under-goes digestion in this sac.
3 Copy and complete the following table:

enzyme	secreted by	region of gut where enzyme acts	substrate	end products
salivary amylase		mouth		maltose
	stomach wall			peptides
lipase		small intestine		
	pancreas		starch	
trypsin			protein and peptides	peptides and amino acids
maltase	wall of small intestine		maltose	
peptidase			peptides	

9 Respiratory exchange of gases

Comparison of inhaled and exhaled air

Oxygen content
The apparatus in figure 9.1 is used to collect a sample of exhaled air. Exhaled air is found to support the combustion of the candle flame for a much shorter time than does an equal volume of inhaled (atmospheric) air. This shows that exhaled air contains less oxygen than inhaled air.

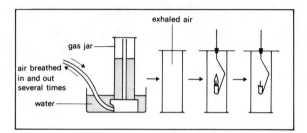

Figure 9.1   Oxygen content of exhaled air

Carbon dioxide content
When air is inhaled through tube X (figure 9.2), it first bubbles through the lime water in A. When air is exhaled through tube X, it passes through the lime water in B before escaping. Lime water B turns much more milky than lime water A showing that exhaled air contains more carbon dioxide (CO_2) than inhaled air.

Water vapour content
Look at figure 9.3. Much more water vapour is found to condense on the outside of flask B (figure 9.3) than on flask A showing that exhaled air contains more water vapour than inhaled air.

Temperature
Inhaled air is colder than exhaled air.

Quantitative analysis of inhaled and exhaled air

An air bubble is drawn into the **J-tube** (figure 9.4) by turning the screw, the apparatus is left in a trough of water for two minutes. The air bubble (now at the temperature of the water) is measured and then **potassium hydroxide** solution (which absorbs CO_2) is drawn in. The bubble is remeasured after a further two minutes in the water trough and then **alkaline pyrogallol** solution (which turns dark brown on

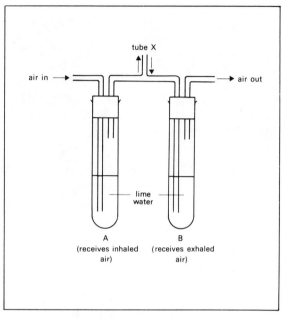

Figure 9.2 CO_2 content of inhaled and exhaled air

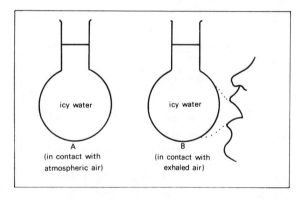

Figure 9.3   Water vapour content

absorbing oxygen) is introduced. The length of the bubble is again measured after a further two minutes. (The latter procedure eliminates any temperature changes which would directly affect the volume of the gas.)

Table 9.1 shows a typical set of results for three different types of air. The CO_2 content of inhaled air is so tiny, however, that it cannot be accurately measured by this method.

During, and immediately after exercise, exhaled air is found to contain more CO_2 and less O_2 than air exhaled normally. Also the rate and depth of breathing increase in order to satisfy the body's demand for more oxygen.

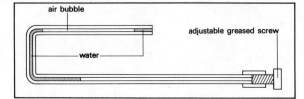

Figure 9.4   J-tube

	inhaled air	exhaled air	exhaled air after exercise
length of original bubble	100 mm	100 mm	100 mm
length of bubble after using potassium hydroxide (CO_2 now removed)	100 mm	96 mm	95 mm
percentage of CO_2	0% (0·03% = correct value)	4% $\left(\frac{100-96}{100} \times 100\right)$	5% $\left(\frac{100-95}{100} \times 100\right)$
length of bubble after using alkaline pyrogallol (O_2 now removed)	80 mm	80 mm	80 mm
percentage of O_2	20% $\left(\frac{100-80}{100} \times 100\right)$	16% $\left(\frac{96-80}{100} \times 100\right)$	15% $\left(\frac{95-80}{100} \times 100\right)$

Table 9.1   J-tube results

Gaseous exchange in other organisms

Green plant in darkness
Look at figure 9.5. Since lime water A remains clear, this indicates that all the CO_2 in the original incoming air has been removed by the sodium hydroxide. Since lime water B turns milky, this shows that CO_2 is produced by a green plant during respiration. The plant is kept in darkness to prevent photosynthesis masking respiration (see chapter 13).

Bicarbonate indicator and CO_2
Look at figure 9.6. After a few hours, the bicarbonate indicator in tubes A, B and C turns from red to yellow (see table 9.2), showing that each organism has given out CO_2 during respiration. Tube D, the control, which remains unchanged shows that the results are

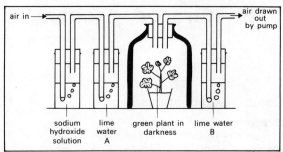

Figure 9.5   Release of CO_2 by green plant

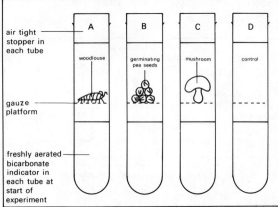

Figure 9.6   Release of CO_2 by living organisms

colour of bicarbonate indicator solution	relative CO_2 concentration
yellow	high (above atmospheric)
red	medium (atmospheric)
purple	low (below atmospheric)

Table 9.2   Bicarbonate indicator range

valid (and not due to some other factor such as the air already present in each test tube).

Universal indicator and pH
Look at figure 9.7. Although the initial colour of the water in the test tube is yellow (see table 9.3), it changes to orange after a few hours. This shows that during respiration, the snails have given out CO_2 which makes the water slightly **acidic**. A control tube is found to remain unchanged.

Gas uptake

Look at figure 9.8. CO_2 given out by the locust is absorbed by the sodium hydroxide. Oxygen taken in by the locust causes the volume of air in the enclosed system to decrease and therefore the drop of liquid moves up the tube.

The volume of air that must be injected (using the syringe) to return the drop of liquid to its original position is equivalent to the volume of oxygen taken up by the locust.

Revision questions

1 Copy and complete the following table:

	exhaled air	inhaled air
oxygen content	less	more
CO_2 content		
water vapour content		

2 Name the 2 chemicals required for a quantitative analysis of exhaled air and for each state which gas it absorbs.

3 Briefly explain why:
(a) In winter, windows, previously clear, steam up once the class has been in the room for an hour.
(b) Air exhaled after exercise turns bicarbonate indicator yellow more rapidly than normal exhaled air.

4 Strictly speaking 2 clips should have been fitted to the apparatus shown in figure 9.2. State where they should have been put and explain why.

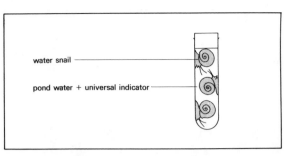

Figure 9.7 CO_2 release and pH

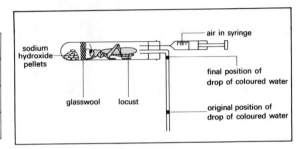

Figure 9.8 Measuring the volume of gas taken up

colour of universal indicator	pH condition
strongly acidic	red
weakly acidic	orange
neutral	yellow
weakly alkaline	green
strongly alkaline	blue

Table 9.3 Universal indicator range

10 Respiratory organs and surfaces

Mammalian respiratory system

The **lungs** (figure 10.1) are a mammal's organs of gaseous exchange. Air entering by the nose or mouth passes via **larynx, trachea, bronchus** and **bronchioles** finally reaching tiny air sacs, the **alveoli**. The alveoli are so numerous that they provide a large surface area for gaseous exchange and give the lungs a sponge-like texture.

The trachea and bronchi, held permanently open by incomplete rings of **cartilage**, are lined with tiny hair-like **cilia** and glandular cells which secrete sticky **mucus**. Rhythmic beating of the cilia carries the mucus, containing trapped dust and germs, upwards to the larynx from where it passes into the gullet.

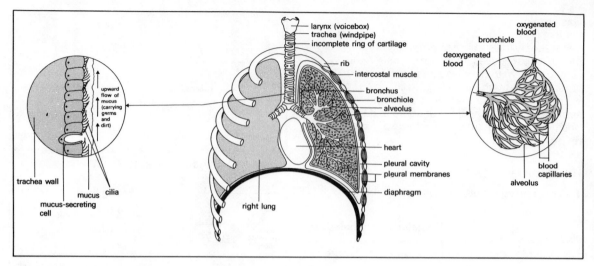

Figure 10.1 Mammalian respiratory system

Breathing movements

Inspiration

Contraction of the **intercostal muscles** pulls the rib cage out and up. At the same time contraction of the **diaphragm** lowers (flattens) the floor of the chest cavity. The volume of the chest cavity is therefore increased (and so pressure decreased) causing air to be inhaled (figure 10.2).

Expiration

On relaxation of the intercostal muscles, the rib cage moves down and in. Relaxation of the diaphragm (back to its dome shape) causes a reduction in volume (and increase in pressure) of the thorax. Air is therefore exhaled (figure 10.2).

The model shown in figure 10.3a demonstrates the action of the diaphragm, but fails to show the action of the intercostal muscles. The reverse is true of the model in figure 10.3b.

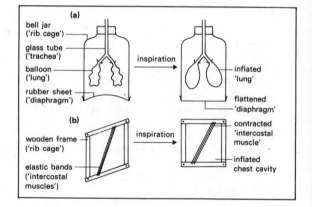

Figure 10.3 Models showing action of (a) diaphragm (b) intercostal muscles

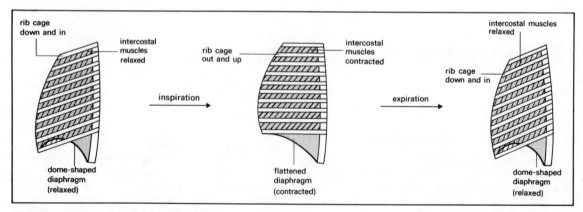

Figure 10.2 Breathing movements

Pleural cavity

Lining the inner walls of the thorax, the upper surface of the diaphragm and the outside of the lungs are the pleural membranes which enclose the **pleural cavity** (figure 10.1). This sac contains lubricating fluid which reduces friction between lungs and thorax during breathing movements.

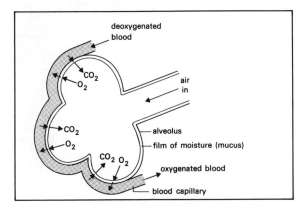

Figure 10.4   Gaseous exchange in an alveolus

Respiratory surface

Oxygen diffuses into the blood because inhaled air contains more oxygen than deoxygenated blood. The oxygen first dissolves in the moisture on the surface of the thin lining of an alveolus (figure 10.4) and then diffuses into the blood in the capillaries. Oxygen combines with haemoglobin (see chapter 16) and is transported to every living cell. CO_2 in deoxygenated blood diffuses out into the alveolus from where it is exhaled.

Respiratory ailments

Tuberculosis is an infectious disease of the lungs caused by a bacterium. The number of deaths due to TB has dropped dramatically in recent years thanks to early detection by X-rays and successful treatment using antibiotics.

    Pneumonia is an infectious disease of the bronchioles which occasionally follows colds or influenza. It is caused by a bacterium and can be cured by antibiotic treatment.

    The cause of **lung cancer**, a malignant growth of lung cells, is unknown though it is aggravated by tobacco smoke and deisel fumes. The number of deaths due to lung cancer is increasing and although it is sometimes successfully cured in its early stages by drugs and surgery in its advanced stages it is incurable.

    Bronchitis is an inflammation of the mucus membrane of the bronchial tubes. It is caused by bacteria which thrive in inhaled air which is damp or

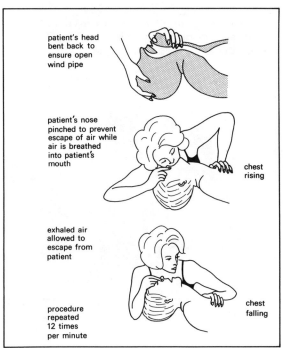

Figure 10.5   Mouth-to-mouth method of artificial respiration

laden with tobacco smoke. It is treated with antibiotics (see chapter 36).

    Pleurisy, a chronic infection and inflammation of the pleural membranes can also be successfully treated with antibiotic drugs.

Artificial respiration

If as a result of a sudden illness or accident (electric shock, drowning etc.) a person's breathing stops, then it must be re-started by artificial respiration (figure 10.5) or the person will die from shortage of oxygen.

Other respiratory surfaces

Gills

A bony fish has four pairs of **gills** (figure 10.6) supported by gill bars and protected on each side by an **operculum** (gill flap). Since each gill is divided into many tiny, thin-walled branches (**filaments**), each containing a rich blood supply, it presents a large surface area to passing water for gaseous exchange.

    Co-ordinated movements of mouth and operculum (figure 10.7) cause water to flow through the gills which absorb oxygen from it. Carbon dioxide diffuses in the opposite direction.

Skin

Earthworms and amphibians absorb O_2 and lose CO_2 directly through their thin moist skin. This is called **cutaneous** respiration (figure 10.8).

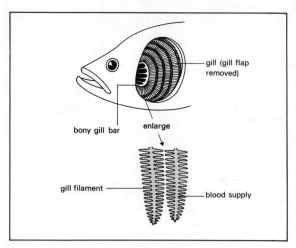

Figure 10.6 Respiratory organs of bony fish

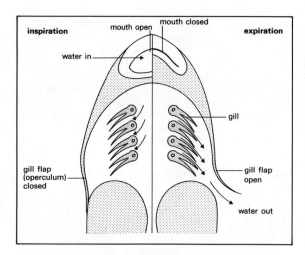

Figure 10.7 Gaseous exchange in a fish

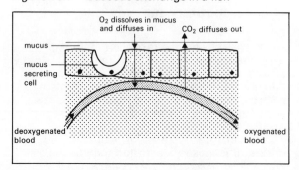

Figure 10.8 Cutaneous respiration

Tracheoles

Tiny holes called **spiracles** on the side of an insect's body allow air to enter and pass along tubes (**tracheae**). These branch into many tiny **tracheoles** (see figure 10.9) which are in direct contact with living tissues. Since oxygen dissolves in the fluid at the end of a tracheole and diffuses directly into the cells, no respiratory pigment is present in an insect's blood. Carbon dioxide diffuses out in the opposite direction.

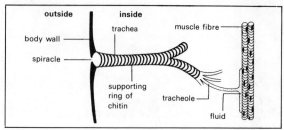

Figure 10.9 Gaseous exchange in an insect

Lenticels

These are tiny holes (figure 10.9) in the stems of plants which allow gaseous exchange to occur by diffusion.

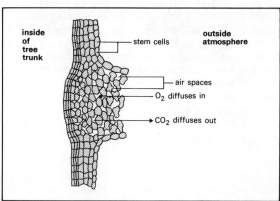

Figure 10.10 Lenticel

Stomata

These are tiny holes which allow gases into and out of green leaves (see figure 18.5).

Characteristics of respiratory surfaces

1 Large surface area to trap oxygen
2 Moist surface to allow oxygen to dissolve
3 Thin surface to allow oxygen to diffuse through easily
4 Blood vessels (present in many animals) to pick up and transport oxygen.

Revision questions

1 Copy and complete the following table using the answers: earthworm, oak leaf, gerbil, oak tree trunk.

respiratory surface	organism
lenticel	
alveolus	
stoma	
skin	

2 (a) Name 4 characteristics of a gill that make it an efficient respiratory surface.
  (b) Which of these characteristics are true of all respiratory surfaces?
3 Copy the following sentence and complete the blanks:
During _____, the diaphragm relaxes and returns to its _____ shape. At the same time the inter-costal muscles _____, the rib cage moves _____ and _____ and air is _____ from the lungs.
4 Which of the following organisms exchanges gases by cutaneous respiration? Toad, herring, pigeon, monkey.

11 Energy release during respiration

Aerobic respiration

Once living cells have received a supply of **glucose**, they must release the **energy** from it. When a sample of glucose is burned in oxygen it gives out its energy all at once as heat (and light). However energy released so quickly cannot be used by living cells. Instead, living organisms release the energy from food in gradual stages, with each step in the long biochemical pathway being controlled by a **respiratory enzyme** located inside the cell. This enzyme-controlled **oxidation** of glucose, often referred to as **tissue respiration**, is summarised in the following equation:

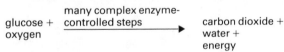

glucose + oxygen $\xrightarrow{\text{many complex enzyme-controlled steps}}$ carbon dioxide + water + energy

Respiration involving the use of **oxygen** is called **aerobic** respiration.

Heat production by germinating seeds
Inversion of the vacuum flask in figure 11.1 ensures that the bulb of the thermometer is in contact with the peas and that the stem is exposed (allowing temperature readings without disturbing the experiment).

After a few days the temperature rises gradually. This shows that during germination pea seeds release heat energy as a result of tissue respiration. A control flask containing boiled and cooled seeds shows no temperature rise in the first few days since

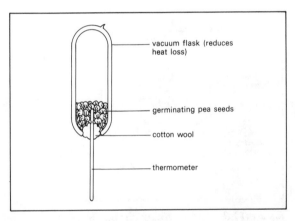

Figure 11.1 Energy release by seeds

the peas are dead. After a few more days however, heat energy is released as saprophytic microbes begin to attack the seeds.

Heat production by animals during respiration
Look at figure 11.2. Although manometer levels X and Y are equal at the start of the experiment, after a few minutes X drops and Y rises. The conclusion is that the animal is giving out heat energy as a result of tissue respiration. This heat, unable to escape, gradually heats up the air trapped in the nearby test tube causing the air to expand and depress level X. Since the only difference between sides A and B is the presence or absence of the respiring animal, side B acts as a control.

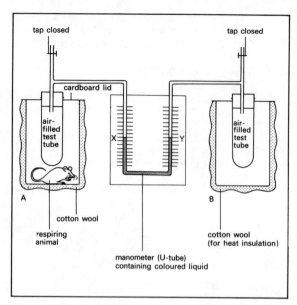

Figure 11.2   Energy release by respiring animal

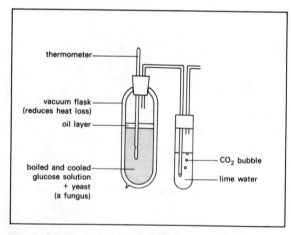

Figure 11.3   Anaerobic respiration in yeast

Anaerobic respiration

Respiration which occurs in the **absence** of oxygen is called **anaerobic** respiration.

Anaerobic respiration in yeast cells

Look at figure 11.3 before setting up the experiment. The glucose solution is boiled to remove dissolved oxygen. The oil layer keeps air, and therefore oxygen, out. Despite the absence of oxygen the yeast cells are found to produce a little heat energy and release bubbles of CO_2 (the clear lime water turns milky). In addition when the yeast and glucose mixture is later distilled at 80°C, the distillate obtained contains **ethanol** (ethyl alcohol). Since the

control experiment using dead yeast cells remains unchanged, it is concluded that the yeast is able to respire anaerobically as summarised in the following equation (which is also referred to as the process of **fermentation**):

glucose ⟶ ethanol + carbon dioxide
+ a little heat energy

Yeast is used in baking and brewing (see chapter 36) and when oxygen is present it can respire aerobically.

Anaerobic respiration in human muscles

Under normal circumstances the energy needed for contraction of muscles comes from aerobic respiration involving glucose and oxygen. However during vigorous activity when muscles are worked continuously, the supply of oxygen temporarily runs out. Under such conditions of oxygen shortage the muscles continue to use glucose and respire anaerobically as follows:

glucose ⟶ lactic acid + a little energy

Thus during intense exertion the concentration of **lactic acid** tends to increase (figure 11.4). However as it reaches intolerable levels it reduces the efficiency of the muscles causing them to fatigue rapidly. Oxygen is required to convert the lactic acid to a non-poisonous state so the muscles can work properly again. Because of this, the body is said to build up an **'oxygen debt'** during anaerobic respiration. In the presence of oxygen this debt is repaid during a rest period, the lactic acid concentration drops and rate of breathing which will have increased dramatically returns to normal.

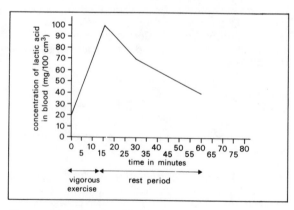

Figure 11.4   Lactic acid graph

Metabolism

Metabolism is the sum of all the chemical changes that occur in a living organism. These include the building up of molecules (e.g. protein) and the

	aerobic	anaerobic
	oxygen always required	oxygen never required
	efficient method of respiration releasing much energy	inefficient method of respiration releasing little energy
	CO_2 and water produced	CO_2 and ethanol produced in plant tissues lactic acid produced in animal tissues and bacteria

Table 11.1 Comparison of aerobic and anaerobic respiration

breakdown of molecules (e.g. glucose). Tissue respiration is therefore an important **metabolic process** which releases energy.

Revision questions

1 **(a)** Using all of the following words, write 1–2 sentences to describe aerobic respiration: enzymes; energy; oxygen; glucose; water; carbon dioxide; oxidation.
  (b) State 2 differences between aerobic and anaerobic respiration in a yeast cell.
2 Suggest why a control using peas soaked in formalin (poisonous preservative) is preferable to the control described for the experiment shown in figure 11.1.
3 **(a)** What gas is absorbed by sodium hydroxide in the experiment in figure 11.5?
  (b) What form of respiration is exhibited by the bean seeds in this experiment?
  (c) What additional end product would be found inside the bean seeds?

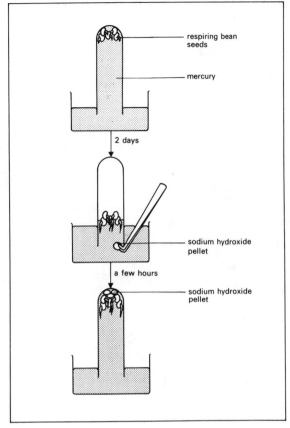

Figure 11.5   see question 3

4 **(a)** If the trend shown in the graph in figure 11.4 continues at the same rate, what concentration of lactic acid will be present in the blood after a further 20 minutes of rest?
  (b) Upon what condition other than rest does this decrease in lactic acid concentration depend?

12 Photosynthesis

Photosynthesis is the process by which green leaves convert **light** energy (trapped by the green pigment chlorophyll) into **chemical** energy (carbohydrate food). **Water** (from the soil) and **carbon dioxide** (from the air) are the two raw materials used up during this synthesis.

Conditions required for photosynthesis

Light
The series of steps shown in figure 12.1 are performed to test a leaf for starch. Discs from a leaf that has been in light are found to turn blue-black where-

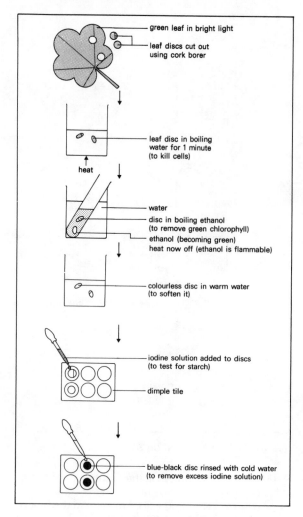

Figure 12.1 Testing a leaf for starch

green leaf in bright light

leaf discs cut out using cork borer

leaf disc in boiling water for 1 minute (to kill cells)

heat

water
disc in boiling ethanol (to remove green chlorophyll)
ethanol (becoming green) heat now off (ethanol is flammable)

colourless disc in warm water (to soften it)

iodine solution added to discs (to test for starch)

dimple tile

blue-black disc rinsed with cold water (to remove excess iodine solution)

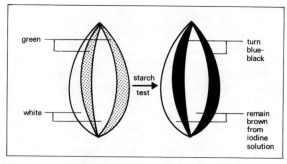

green

turn blue-black

white

starch test

remain brown from iodine solution

Figure 12.2 Need for chlorophyll

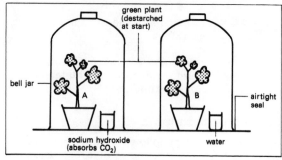

green plant (destarched at start)

bell jar

A

B

airtight seal

sodium hydroxide (absorbs CO_2)

water

Figure 12.3 Need for CO_2

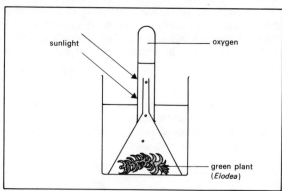

sunlight

oxygen

green plant (*Elodea*)

Figure 12.4 Oxygen release

as those from a leaf in darkness do not react with brown iodine solution. It is concluded therefore that **light** is necessary for photosynthesis.

Chlorophyll
When the procedure is repeated with a leaf that is **variegated** (i.e. has two or more colours one of which is green), only the green regions give a positive result with iodine solution (see figure 12.2) showing that green **chlorophyll** is necessary for photosynthesis.

Water
Water is also essential for photosynthesis.

Carbon dioxide
Before being used in the experiment shown in figure 12.3, the two plants are kept in darkness for two days to ensure that at the start of the experiment they

contain no starch. In the experiment they are left under the conditions shown in figure 12.3 for two days and then leaf discs from each plant are tested for starch. Since those from A fail to give a positive result, but those from B turn blue-black, it is concluded that **carbon dioxide** is essential for photosynthesis.

By-product of photosynthesis
The gas given off by the waterweed shown in figure 12.4 is found to relight a glowing splint. This shows that, in addition to food, **oxygen** is produced during photosynthesis.

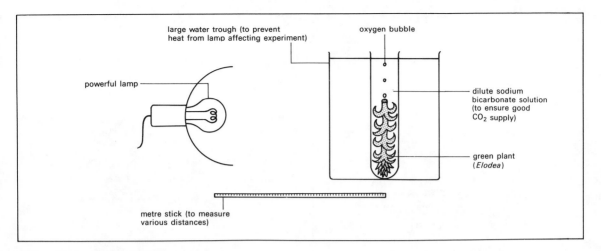

Figure 12.5   Elodea bubbler experiment

Elodea bubbler experiment

Figure 12.5 shows an investigation into the effect of light intensity on photosynthetic rate. The number of oxygen bubbles released per minute by the cut end of an *Elodea* stem indicates the rate at which photosynthesis is proceeding. At first the lamp is placed exactly 100cm from the plant and the number of oxygen bubbles released per minute counted. The lamp is then moved to a new position (say 60cm from the plant) and the rate of bubbling noted (once the plant has had a short time to become acclimatized to this new higher light intensity). The process is repeated for lamp positions even nearer the plant as shown in table 12.1. When this typical set of results is displayed as a graph (figure 12.6), it can be seen that as light intensity increases, photosynthetic rate also increases until it reaches a maximum of 25 bubbles per minute at around 64 units of light.

Limiting factor
Further increase in light intensity does not increase the photosynthetic rate. This is because shortage of CO_2 is now holding up the process. CO_2 concentration is therefore said to be the **limiting factor** and when more CO_2 is supplied (as sodium bicarbonate solution), the rate of bubbling increases again.

Effect of temperature
The graph in figure 12.7 shows that the photosynthetic rate rises to an optimum at around 40°C and then rapidly drops. This is because photosynthesis consists of many reactions controlled by **enzymes** which are destroyed at high temperatures.

Sugar-starch interconversions

When de-starched leaf discs have been floated on glucose solution in darkness for a few days, they are

distance from plant (cm)	units of light (calculated using mathematical formula)	number of oxygen bubbles/ per min
100	4	4
60	11	10
40	25	19
30	45	24
25	64	25
20	100	25

Table 12.1   Elodea bubbler results

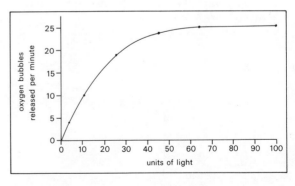

Figure 12.6   Graph of *Elodea* bubbler results

found to turn blue-black on being tested with iodine solution. This shows that green leaves can build up glucose molecules (normally made by photosynthesis) into starch.

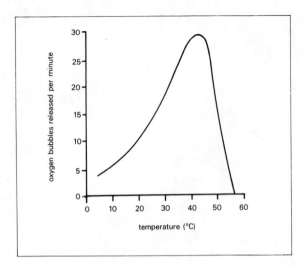

Figure 12.7 Effect of temperature on photo-synthesis

The mechanism of this synthesis is investigated in the following experiment. A sample of potato extract is prepared by liquidising a mixture of fresh potato tuber and water and then centrifuging the mixture until the **supernatant** (figure 12.8) is starch-free. This '**potato extract**' is added to an active form of glucose (**glucose-1-phosphate**) in each of four dimples in row A of a tile (figure 12.9). Rows B and C are controls. One dimple for each condition is tested with iodine solution at 3 minute intervals and starch is found to be formed in row A only.

An enzyme in the potato extract has promoted the synthesis of glucose (the substrate) to starch (the product) as shown in figure 12.10. The enzyme needed to do this, **starch phosphorylase**, is also found in other regions of the plant, such as leaf cells, where it converts the excess glucose formed during photosynthesis to starch.

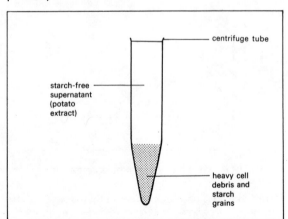

Figure 12.8 Preparation of potato extract

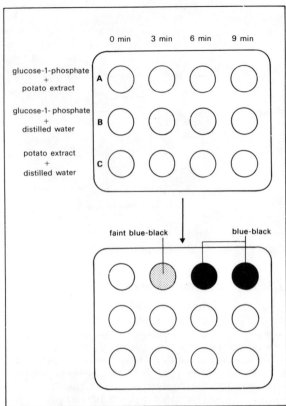

Figure 12.9 Conversion of sugar to starch

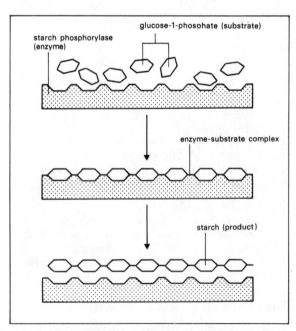

Figure 12.10 Action of starch phosphorylase

Equation of photosynthesis

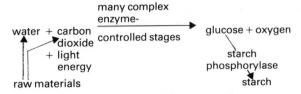

water + carbon dioxide + light energy

$\xrightarrow[\text{controlled stages}]{\text{many complex enzyme-}}$ glucose + oxygen

raw materials

starch $\xrightarrow{\text{phosphorylase}}$ starch

Organ of photosynthesis

A leaf (figure 12.11) is well adapted to suit its function of photosynthesis. Its wide flat shape presents a **large surface area** of both **chlorophyll** (for the absorption of solar energy from above) and **stomata** (for gaseous exchange with the air below). Since most chloroplasts are found in the **palisade mesophyll** layer, it is the chief photosynthetic region of the leaf; however the spongy mesophyll beneath it also photosynthesises to a lesser extent. Each **stoma** is surrounded by two **guard** cells which make it close in darkness and open in light thus controlling **gaseous exchange** (entry of CO_2 and exit of oxygen and water vapour).

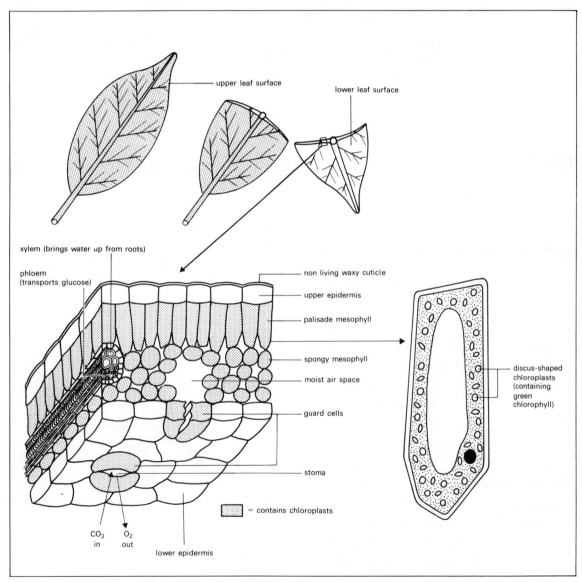

Figure 12.11 Structure of leaf (from dicotyledonous plant)

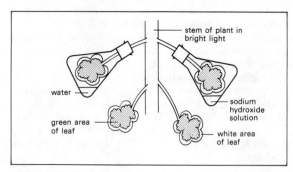

Figure 12.12   see question 2

Revision questions

1 Write 1–2 sentences to describe the process of photosynthesis including all of the following words in your answer: chlorophyll; oxygen; chemical energy; water; enzymes; light energy; glucose; carbon dioxide.

2 **(a)** Several factors are essential for photosynthesis to occur. Which 2 could be demonstrated using the apparatus shown in figure 12.12?

   (b) Why is there no need for a second plant as a control in this experiment?

   (c) Testing a leaf for starch involves the following steps: boiling in alcohol; rinsing in cold water; adding iodine solution; boiling in water; rinsing in warm water. List these in the correct order and opposite each state the reason for carrying out the step.

3 In an *Elodea* bubbler experiment, it was found that a small increase in CO_2 concentration increased the rate of bubbling, but that higher concentrations of CO_2 failed to bring about further increase. What factor was limiting?

13   Balance between photosynthesis and respiration

Plants and carbon dioxide

Plants respire 24 hours per day, but only photosynthesise when light is present. When the four test tubes shown in figure 13.1 were left in light for several hours, the results given in table 13.1 were obtained. The contents of tube A turn yellow since CO_2 has been released by the plant respiring in darkness. The contents of tube B remain unaltered since the amount of CO_2 formed by respiration exactly equals the amount of CO_2 used up during photosynthesis (which proceeds at a very low rate owing to the dim light). In bright light photosynthesis occurs at a rate many times faster than respiration. Thus the contents of tube C turn purple since much CO_2 is taken in by the plant for rapid photosynthesis (which masks the small respiratory output of CO_2). The relationship between CO_2 and a green leaf is summarised in figure 13.2.

Compensation point
The light intensity at which the rate of CO_2 uptake by a plant for photosynthesis exactly balances the rate of CO_2 production by the plant during respiration, is called the **compensation point**. During 24 hours of clear weather, this occurs twice – just after dawn and just before nightfall. Between these two compensa-

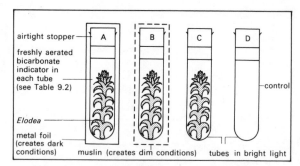

Figure 13.1   Plants and CO_2

	A	B	C	D
initial colour of indicator	red	red	red	red
final colour of indicator	yellow	red	purple	red

Table 13.1   Results of experiment in figure 13.1

tion points in "daytime", photosynthesis by far exceeds respiration and a store of carbohydrate builds up. Some of this is used by the plant during the night and on dull days.

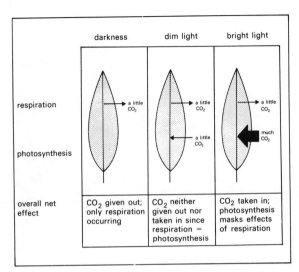

Figure 13.2   Output and intake of CO_2

tube C represents the **balance of nature**. Animals depend on plants for oxygen and food.

respiration	photosynthesis
occurs in all living cells	only occurs in plant cells containing green chlorophyll
occurs 24 hours per day	only occurs in presence of light
oxygen used up	oxygen produced
CO_2 produced	CO_2 used up
food used up (plant loses weight)	food produced (plant gains weight)
energy given out	energy taken in

Table 13.2   Comparison of respiration and photosynthesis

Plants, animals and CO_2

When the four test tubes shown in figure 13.3 are left in light for a few hours, the results given below each one are obtained. CO_2 intake by the photosynthesising plant in tube A causes the indicator to turn purple as before. CO_2 output by the respiring animal in tube B causes the indicator to turn yellow. In tube C the indicator remains red because the CO_2 taken in by the plant for photosynthesis equals the CO_2 given out by the animal (and plant) during respiration. Similarly the oxygen given out by the plant as a result of photosynthesis equals that taken in by the animal (and plant) for respiration. Thus

Revision questions

1 Strictly speaking what two additional controls should have been included in the experiment shown in figure 13.1?
2 Addition of CO_2 to water lowers the pH, removal of CO_2 raises the pH.
   (a) Which tube in figure 13.1 will have the lowest pH?
   (b) Which tube in figure 13.3 will have the highest pH?
   (c) Explain your choice in each case.
3 (a) Explain why the indicator in the tube shown in figure 13.4 turns from red to yellow after several hours in light.

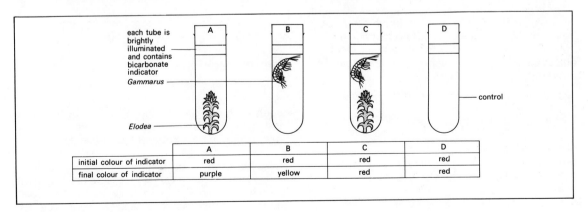

	A	B	C	D
initial colour of indicator	red	red	red	red
final colour of indicator	purple	yellow	red	red

Figure 13.3   Plants, animals and CO_2

41

(b) Without altering the contents of the tube how could the experiment be adapted to make the indicator return to a red colour?

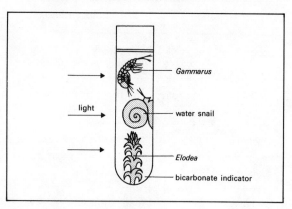

Figure 13.4 see question 3

4 Copy figure 13.5 and complete it using the words: respiration; water; photosynthesis; oxygen.

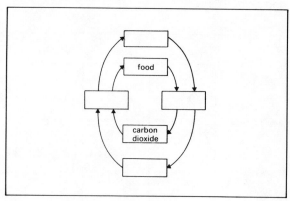

Figure 13.5 see question 4

14 Food chains

Every living thing is part of a biological system called the **ecosystem**. The energy required by all living things is put into its biological system by **producers** (green plants) which fix solar energy during photosynthesis and convert it into chemical energy contained in food. The energy in this food is transferred to a herbivore (**primary consumer**) when it eats the plant producer, then to a carnivore (**secondary consumer**) when it eats the herbivore and so on through a series of organisms. Such a relationship where one organism feeds on the previous one and in turn provides food for the next one in the series is called a **food chain** as shown in the following example. Each arrow represents flow of energy.

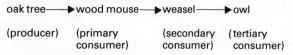

oak tree	wood mouse	weasel	owl
(producer)	(primary consumer)	(secondary consumer)	(tertiary consumer)

However such a food chain rarely occurs in isolation. Normally the producer is eaten by several herbivores which are in turn preyed upon by several different predators. Such a system of interconnecting food chains is called a **food web** (figures 14.1 and 2).

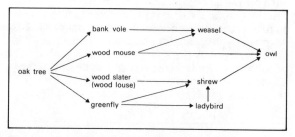

Figure 14.1 Woodland food web

Energy transfer

As energy flows along a food chain, a progressive **loss** occurs for two reasons. Firstly, an organism expends energy on building its body. However this may include parts such as cellulose cell walls or bone or skin or horns etc. which when eaten by the succeeding consumer have little nutritional value. These parts tend to be left uneaten or to be expelled, undigested, as faeces and as a result, energy is lost to the food chain. (Some of this energy is however gained by the ecosystem's decomposers – see chapter 33).

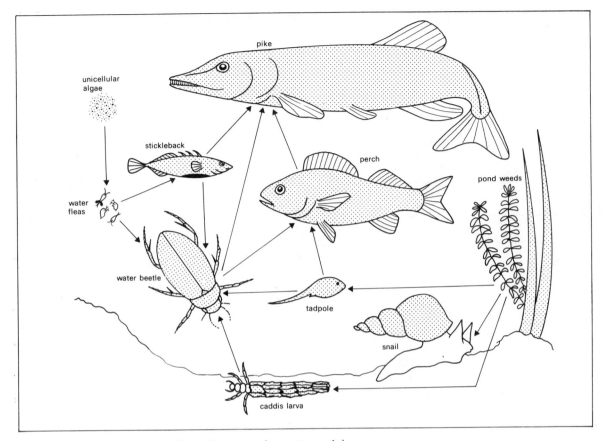

Figure 14.2 Pond food web (organisms not drawn to scale)

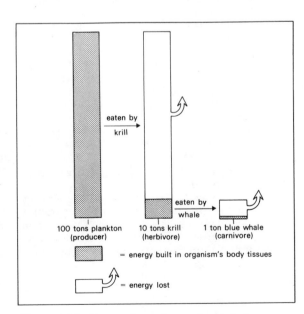

Figure 14.3 Energy loss along a food chain

Secondly, most of the energy gained by the consumer in its food is used for keeping warm and moving about. A lot of energy is therefore lost as heat and only about 10% of the energy taken in by an organism is incorporated into its body tissues. Figure 14.3 shows the fate of energy as it is transferred along a food chain. More efficient use is made of plants by humans consuming them directly rather than first converting them into animal products since this cuts out one of these energy-losing stages in the food chain.

Pyramid of numbers and biomass

The total number of individuals at each level in a food chain decreases. This is called a **pyramid of numbers** (figure 14.4) and occurs for two reasons. Firstly the energy loss that occurs at each link limits the amount of living matter that can be supported at the next level. Secondly the final consumer tends to be larger than its predecessors.

The **biomass** of a population is its total weight of living matter. In a food chain, the biomass decreases at each level and therefore also takes the form of a pyramid.

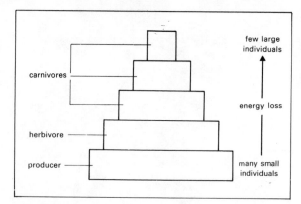

Figure 14.4   Pyramid of numbers

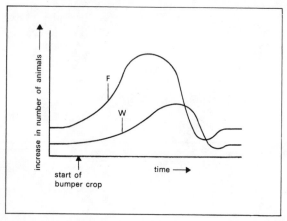

Figure 14.5   The balance disturbed

Removal of the carnivore

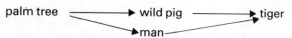

When tigers were exterminated from the above food web by man, the wild pigs quickly increased in number and consumed all the young palm trees thus depriving man of future food and shelter.

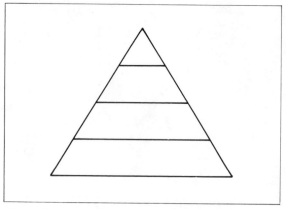

Figure 14.6   see question 1

Disturbing the balance

In a balanced food web, a certain amount of plant material grows continuously and supports a fairly constant number of herbivores which are in turn consumed by a fairly constant number of carnivores and so on. However this balance (**equilibrium**) is disturbed if the numbers of one of the organisms in the food web changes drastically.

Over-abundance of the producer
The graph in figure 14.5 shows the effect of a bumper wheat crop on the F (field mouse) and W (weasel) populations. As the number of field mice increases so also does the number of weasels. The intense predation that follows reduces the mouse population and eventually many weasels cannot find food. Thus the decrease in number of prey (F) is reflected in the number of predators (W) and eventually the original equilibrium is restored.

Revision questions

1 (a) Copy the pyramid of numbers shown in figure 14.6 and complete it using the organisms: waterflea, pike, algae, stickleback.
  (b) Which of these organisms is the secondary consumer?
  (c) Which population of organisms in this pyramid contains the most potential energy?
  (d) Compared to the other organisms, what rule applies to the individual body size of the organisms occupying the top position in a food pyramid?
  (e) Why do the numbers decrease towards the top of a food pyramid?
2 Briefly explain why:
  (a) a pound of meat is always more expensive than a pound of flour.
  (b) corn is converted into chicken more efficiently when the birds are kept in a warm temperature.
3 In 1954/5 when almost all of the rabbits in Britain died of myxomatosis, the following food web was grossly disturbed. Suggest one advantage and one disadvantage to the farmer that resulted.

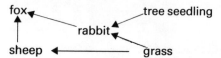

15 Water

Drying experiments

90 °C.

When 100 g portions of various materials of plant and animal origin are slowly dried in an oven to **constant mass**, the percentage loss in mass for each equals its percentage water content. This varies from organism to organism (e.g. 10% for 'dry' seeds, 99% for jellyfish) but they all contain some water since water is the main constituent of protoplasm.

Biological importance

Water is the ideal medium in which: gases can dissolve and enter plant and animal bodies; food can be digested by enzymes; materials can be transported round the organism and soluble wastes can leave an animal's body. In addition water acts as: a means of support (especially in plants); a lubricating agent (pleural cavity); a cooling agent (sweating in animals, transpiration in plants) and a medium in which sperm can swim to and fertilise eggs. Water is also an essential raw material in photosynthesis.

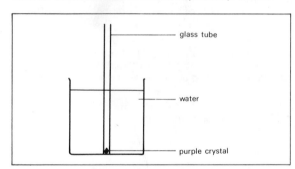

Figure 15.1 Diffusion

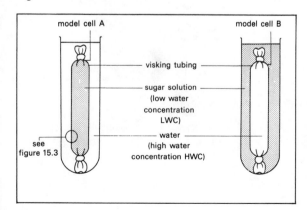

Figure 15.2 Osmosis in model cells

Diffusion

This is the movement of particles of a substance from a region of **high concentration** of that substance to a region of **low concentration** until the concentration becomes uniform.

Look at figure 15.1. When a purple crystal is added through the glass tube to the water, the purple particles diffuse from a region of high concentration up into the water (a region of low concentration) and the water molecules diffuse in the opposite direction until the concentration of purple particles and water is uniform throughout the beaker.

Diffusion of materials occurs in every living cell. For example food and oxygen diffuse from human blood into all the living cells and carbon dioxide diffuses out of the cells into the blood.

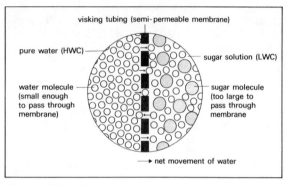

Figure 15.3 Osmosis

Osmosis

Look at figure 15.2. Since model cell A gains weight after 30 minutes and model cell B loses weight, but no sugar passes into the water, it is concluded that the tiny water molecules, but not the larger sugar (solute) particles, are able to pass through the visking tubing. Visking tubing is therefore said to be a **semi- (selectively) permeable membrane** (see figure 15.3).

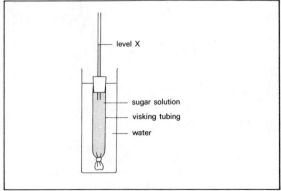

Figure 15.4 Osmometer

Osmosis is the name given to this movement of water molecules from a region of **high water concentration** (low solute concentration) to a region of **low water concentration** (high solute concentration) through a semi-permeable membrane.

The apparatus shown in figure 15.4 is called an **osmometer**. After a few minutes the sugar solution gains water by osmosis and causes level X to rise. The sugar solution is said therefore to exert an **osmotic pressure**. **Osmotic potential (OP)** is the potential of a solution to gain water by osmosis and exert an osmotic pressure if given suitable conditions.

Water relations of cells

Since a cell membrane is semi-permeable and the cell contains solutes and therefore has an osmotic potential (OP), water may pass into or out of the cell by osmosis. The direction depends on the concentration of the liquid in which the cell is immersed.

Red blood cells

Since pure water has a higher water concentration than the contents of red blood cells (figure 15.5), water enters by osmosis until the cells burst. Since 0.85% salt solution has the same water concentration as the cell contents, there is no net flow of water into or out of the cell by osmosis. Since 1.7% salt solution has a lower water concentration than the cells, water passes out and the cells shrink.

Amoeba

Unicellular animals that live in fresh water take in

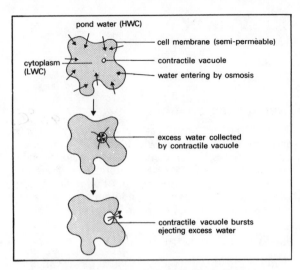

Figure 15.6 Osmoregulation in *Amoeba*

water continuously by osmosis. Bursting is prevented by the **contractile vacuole** (figure 15.6) removing excess water. Such control of water balance is called **osmoregulation**.

Plant cells

Since pure water has a higher water concentration than the contents of a normal plant cell (figure 15.7), water enters the cell by osmosis. The vacuole swells up and presses the cytoplasm against the cell wall which stretches slightly and presses back preventing the cell from bursting. Cells in this swollen condition are said to be **turgid**. Similarly a cylinder of freshly cut potato tissue becomes rigid (turgid) when immersed in water. A young plant depends on this turgidity of its cells to support it.

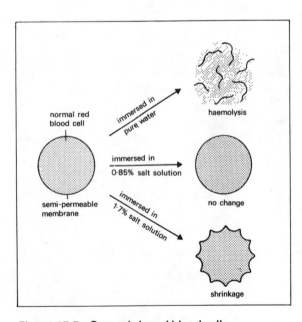

Figure 15.5 Osmosis in red blood cells

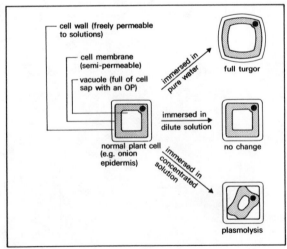

Figure 15.7 Osmosis in plant cells

Since the weak sucrose solution (in figure 15.7) has the same water concentration as the cell contents, no net flow of water occurs.

Since concentrated sucrose solution has a lower water concentration than the cell contents, water passes out of the cell by osmosis. The living contents shrink and pull away from the fairly rigid cell wall. Cells in this state are said to be **plasmolysed**. Similarly a cylinder of freshly cut potato becomes soft (**flaccid**) when immersed in concentrated sucrose solution. If a plant's cells become flaccid, the plant loses support and wilts. Plasmolysed cells are not dead. When immersed in water they undergo **deplasmolysis** (i.e. regain turgor by taking in water by osmosis).

Revision questions

1 Copy and complete the following table:

important role of water	where role is played
agent of support cooling agent raw material lubricating agent agent of transport	

2 Define the terms diffusion and osmosis.
3 Explain why a red blood cell bursts when placed in water but an onion leaf epidermal cell does not.
4 (a) Which cell in figure 15.8 is turgid?
   (b) Which cell is plasmolysed?
   (c) What would cause process X?
   (d) What would cause process Y?
   (e) Which of these processes is called deplasmolysis?

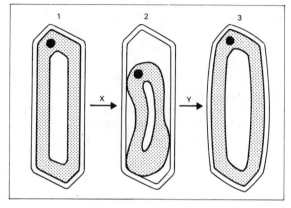

Figure 15.8   see question 4

16   Transport in animals

A tiny unicellular animal (e.g. *Amoeba*) is able to take in sufficient oxygen by **diffusion** through its cell membrane because it presents a relatively large surface area to the surrounding environment. Since a large multicellular animal has a small surface area/volume ratio (see chapter 19), it needs **additional absorbing areas** to take in oxygen and food. In human beings, for example, alveoli in the lungs and villi in the ileum greatly increase the surface area for absorption.

Once essential substances have entered a living organism, they must spread through its body since all of its protoplasm requires oxygen and food. In tiny *Amoeba*, these materials simply diffuse through the cell. However diffusion is too slow for larger organisms where the absorbing organ may be a considerable distance from the waiting cells. In order to carry materials round the body at a rate faster than is possible by diffusion, most organisms have a **transport (mass flow) system**.

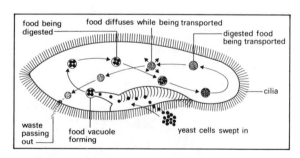

Figure 16.1   Cyclosis in *Paramecium*

Transport inside one cell

Paramecium
Figure 16.1 shows this unicellular animal feeding on yeast cells stained with Congo red (a pH indicator). Movement of the cytoplasm causes the food vacuoles to be circulated round the body in a definite route.

During this process, called **cyclosis**, the stained yeast cells change from red to blue as conditions of low pH (suitable for the action of digestive enzymes) occur. By means of this mass flow system digested food is transported to all parts of the cell at a rate faster than diffusion.

Elodea

Movement of cytoplasm also occurs inside plant cells. In a leaf cell of *Elodea*, **streaming** of the cytoplasm is indicated by the movement of chloroplasts around the inner surface of the cell wall. Such movements make transfer of materials within a cell more efficient.

Transport from cell to cell

Open transport system

In such a system (figure 16.2), the blood is enclosed (e.g. in a tubular heart) for only part of its circuit round the body. During the rest of its journey, it flows along irregular channels bathing the living tissues **directly**. An **open transport system** is found in arthropods such as insects and crustaceans.

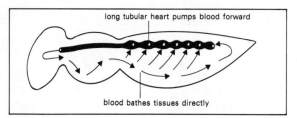

Figure 16.2   Open transport system in an insect

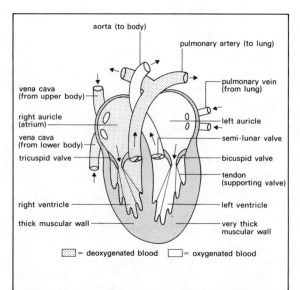

Figure 16.3   Mammalian heart

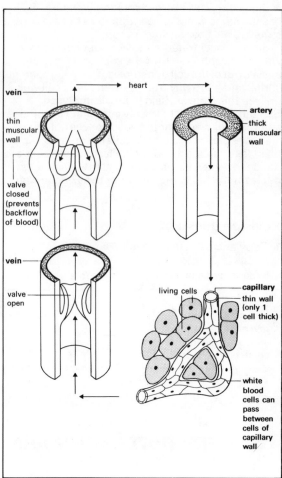

Figure 16.4   Types of blood vessel

Closed transport system

This time the blood is always **enclosed** in tubes called **blood vessels**. It is from **capillaries**, the narrowest of these vessels, that materials are exchanged with living cells. Blood is kept flowing along the vessels by the pumping action of a muscular **heart**. A **closed transport system** is found in earthworms and all vertebrates.

Mammalian heart (figure 16.3)

Deoxygenated blood from the body passes into the **right auricle** by the **venae cavae** and then down into the **right ventricle** through the open **tricuspid valve**. When the muscular ventricle wall contracts, the blood, under pressure, closes the tricuspid valve (thus preventing **backflow**) and rushes out through the **pulmonary artery** to the lungs where gaseous exchange occurs (see chapter 10).

Oxygenated blood returns to the **left auricle** by the **pulmonary veins** and passes through the **bicuspid valve** into the **left ventricle** from where it is forced out through the **aorta** to all parts of the body.

Semi-lunar valves at the entrance to the pulmonary artery and aorta also prevent backflow of blood. The muscular wall of the left ventricle is thicker than that of the right, since the former has to pump blood all round the body whereas the latter only to the lungs. The ventricles contract simultaneously producing **heartbeat** about 70 times per minute when at rest, but about 100 or more during activity or excitement. Compared with an unfit person, a physically fit person's pulse rate does not increase as much during exercise and returns to normal more quickly afterwards.

Vessels (figure 16.4)

An **artery** is a vessel which carries blood away from the heart. It has a thick muscular wall to withstand the **high pressure** of **oxygenated** blood spurting along it. (The pulmonary artery is exceptional in carrying deoxygenated blood). An artery divides into smaller vessels and finally into a dense network of tiny thin-walled **capillaries** which are in close contact with the living cells. Oxygen and food diffuse into the cells and CO_2 and wastes return to the capillaries which unite to form larger vessels which in turn form **veins**. A vein is a vessel which carries blood back to the heart. Although muscular, its wall is thinner than that of an artery since **deoxygenated** blood at **low pressure** oozes along it. Valves are present in veins to prevent backflow. (The pulmonary vein is exceptional in carrying oxygenated blood). A simplified version of man's closed circulatory system is shown in figure 16.5.

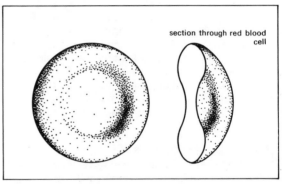

Figure 16.6 Red blood cell

Composition of blood

Red blood corpuscles (figure 16.6)

These contain no nucleus and only last for about three months. New ones are made constantly in the bone marrow. Their **biconcave** disc shape offers maximum surface area for oxygen uptake by **haemoglobin** (red-coloured protein found in their cytoplasm which contains **iron**). The oxy-haemoglobin formed is unstable and releases its oxygen on arriving in capillaries at body cells as follows:

$$\text{haemoglobin} + \text{oxygen} \underset{\text{at body cells}}{\overset{\text{in lungs}}{\rightleftharpoons}} \text{oxy-haemoglobin}$$
(dark red)   (bright red)

Oxygen-deficient habitats
Some invertebrates which live in mud at the bottom of a river (e.g. redworm, chapter 37), have a very rich supply of haemoglobin dissolved in their blood. This enables them to make the most efficient use of any

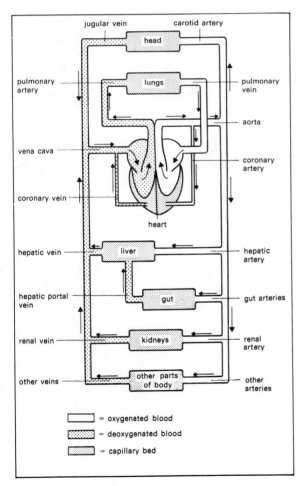

Figure 16.5   Closed blood transport system in man

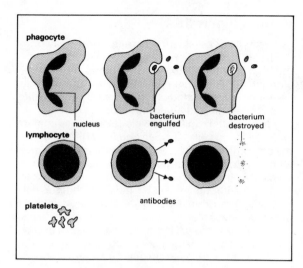

Figure 16.7   White blood cells and platelets

small amount of oxygen present in their oxygen-deficient habitat. Some other invertebrates (e.g. water flea) are normally fairly colourless and only develop an abundant supply of red haemoglobin in poorly aerated water.

Since there is less oxygen in the air at high altitudes, people living in such places are found to have more red blood cells than people living in lowland areas.

White blood cells (figure 16.7)
There are fewer white cells than red cells in human blood (1 white: 600 red) and white cells do contain a nucleus. **Phagocytes** (white cells) made in the bone marrow, engulf bacteria by **phagocytosis**. Since white cells can squeeze between the cells of capillary walls and gather at wounds where they kill disease-causing bacteria, these white blood cells form one of the body's main defence systems. **Lymphocytes**, (another kind of white cell) made in the lymphatic system, produce chemicals called **antibodies** which also destroy microbes. **Pus** is a collection of dead white cells at a wound.

Platelets (figure 16.7)
These tiny cell fragments play an important part in the clotting of blood. At a cut, platelets and damaged tissue release an **enzyme** which promotes the following conversion:

$$\text{soluble fibrinogen} \xrightarrow{\text{enzyme (thrombin)}} \text{insoluble fibrin}$$

Threads of fibrin form a network which traps red blood cells. As this clot dries it forms a barrier over the wound preventing loss of blood and entry of germs.

Plasma
This watery yellow liquid, in which blood cells are suspended, contains many dissolved substances such as glucose, amino acids, minerals, hormones, urea and CO_2 (though much of the CO_2 is returned to the lungs by the red blood cells).

Lymphatic system

Since blood entering a capillary (figures 16.4 and 16.8) is under pressure, much blood plasma is squeezed through tiny pores into the space between the living cells. Useful substances diffuse from this tissue fluid (**lymph**) into the cells and wastes diffuse into the fluid. Since blood in the capillary contains more protein than lymph, some water passes from lymph (High Water Concentration) to blood (Low Water Concentration) by osmosis. As a result of body movements and pressure of accumulation, the remaining lymph drains into tiny, blindly-ending, thin-walled **lymph vessels**. These join up to form the **lymphatic system** (see figure 16.9 which only shows a few vessels). Lymph finally joins the blood again by two **ducts** into the arm veins. Scattered throughout the lymphatic system are **lymph glands** which make lymphocytes and contain cells which engulf passing germs by phagocytosis. Since lymph bathes the cells directly and is not always confined to tubes, the lymphatic system is an **open** transport system.

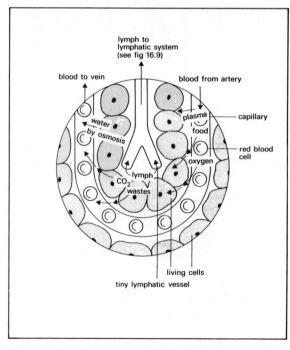

Figure 16.8   Exchange of materials in capillary bed

Revision questions

1 (a) Which has the larger surface area/volume ratio, a frog or *Paramecium*?
(b) Which of these animals requires a further oxygen-absorbing surface in addition to its body's thin outer layer?
(c) Name the corresponding oxygen-absorbing surface in a fish.

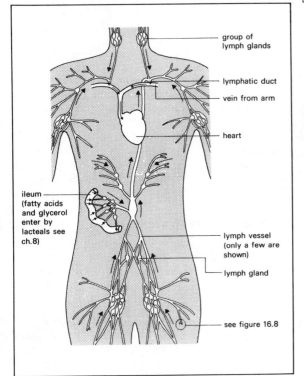

Figure 16.9 Lymphatic system

2 Copy and complete the following:

	red corpuscle	white corpuscle
relative number		
shape		
nucleus present or absent		
function		

3 Write a sentence to explain each of the following statements:
(a) Blood is dark red in a pulmonary artery and bright red in a pulmonary vein.
(b) Swollen neck glands are often suffered during severe illness.
(c) The warmer the water, the redder the population of *Daphnia* (water flea) becomes.

4 (a) Name blood vessels labelled 1, 2 and 3 in figure 16.10.
(b) From where has the blood in vessel 1 come?
(c) Why is a blood clot, occurring at either position A or B, of particular danger to the person?
(d) Explain why the heart wall is thicker at region Y than at region Z.

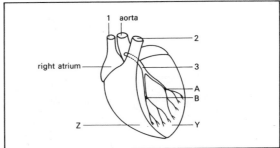

Figure 16.10 see question 4

17 Functions of kidney

Nitrogenous excretion

Excretion is the extraction and elimination from the body of the **waste** products of metabolism. Nitrogenous wastes (e.g. urea) are removed from the blood by the **kidneys** (figures 17.1 and 17.2). Functional units of a kidney are called **nephrons**. Each nephron (figure 17.3) consists of a cup-shaped **Bowman's capsule** leading into a long coiled **tubule**. The renal artery supplying each kidney with blood divides into many small branches each of which leads to a **glomerulus**, a tiny knot of blood capillaries, surrounded by a Bowman's capsule.

Filtration

Since the blood vessel entering a glomerulus is wider than the vessel leaving it, the blood in a glomerulus is under pressure. As a result, plasma fluid filters out through pores in the capillary walls and collects in the Bowman's capsule. This **glomerular filtrate** contains glucose, salts, urea and water but not plasma proteins or blood cells which are too large to pass through the capillary wall.

Reabsorption

As glomerular filtrate passes through a kidney tubule, useful substances (all glucose, some salts and some water) are reabsorbed into the branching network of capillaries surrounding the coiled tubule. Further water is absorbed from the liquid flowing down the collecting duct finally leaving **urine**, which contains all the urea, excess salts and water.

Table 17.1 compares the composition of glomerular filtrate with urine. To bring about such **selective reabsorption**, kidney cells require energy. This is generated during tissue respiration in the kidney cells using oxygen. Blood leaving each kidney by the renal vein is therefore deoxygenated.

substance	glomerular filtrate	final urine
glucose	0.1%	0%
salts (sodium, calcium etc)	1.0%	1.8%
urea	0.02%	2.0%
water	98.5%	96%

Table 17.1 Comparison of glomerular filtrate and urine

Elimination

Urine drains down the **ureters** into the **bladder** from where it is expelled via the urethra by voluntary action of the sphincter muscle.

Control of water balance

Man loses water in urine, sweat, faeces and exhaled air and gains water from food, drink and metabolic water. The latter is the water produced during respiration in living cells:

glucose + oxygen ⟶ **water** + carbon dioxide + energy

If a lot of water is lost from the body then the kidneys produce very concentrated urine thus conserving water; if there is a lot of water in the blood then they make very dilute urine thus ridding the body of excess water. This control of water balance is called **osmoregulation** (see chapter 25 for further details).

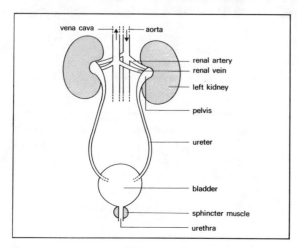

Figure 17.1 Organs of nitrogenous excretion

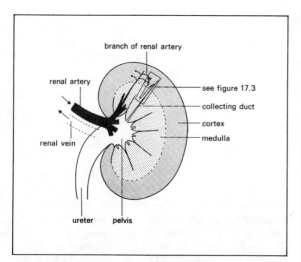

Figure 17.2 Structure of a kidney

Revision questions

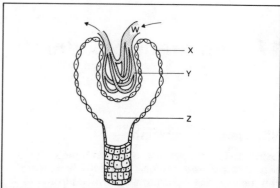

Figure 17.4 see question 1

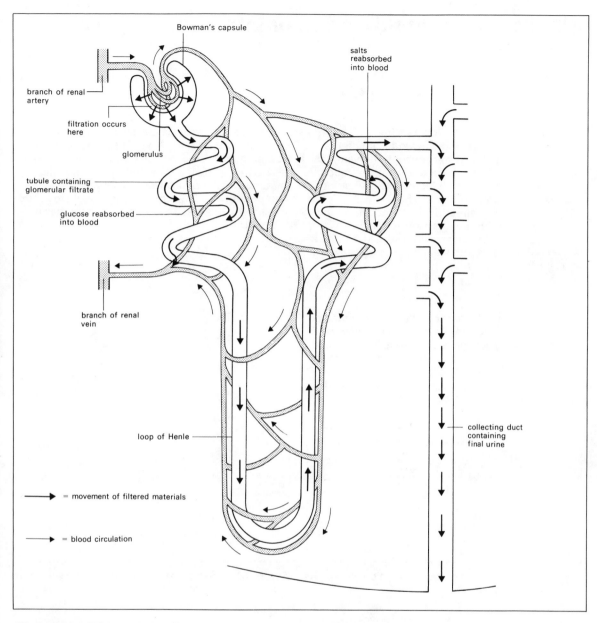

Figure 17.3 Kidney nephron

1 (a) Which vessel supplies the blood arriving at W in figure 17.4 which shows part of a nephron.
   (b) Name structures X and Y.
   (c) What nitrogenous waste passes into Z?
   (d) Using the terms urethra, ureter, kidney tubule, bladder and collecting duct, give the correct route taken by this waste product.
   (e) 180 litres of glomerular filtrate and 1.5 litres of urine are produced on average by man per day. What happens to most of the water in the glomerular filtrate?

2 Explain why blood in a renal vein contains less glucose, less oxygen and more CO_2 than blood in a renal artery.

3 (a) What is meant by the term osmoregulation?
   (b) After excessive sweating, which of the following processes brings about osmoregulation?
   A filtration of blood
   B reabsorption of water
   C elimination of urine
   D excretion of faeces

18 Transport in flowering plants

Role of root

The structure of a root is shown in figure 18.1. Since the cell sap of a **root hair** (an extended epidermal cell) is a region of low water concentration and the soil solution a region of high water concentration, water passes into a root hair (figure 18.2) by osmosis. Water then passes from the root hair (which now has a higher water concentration) into a neighbouring **cortex cell** (which, by comparison, has a lower water concentration) and so on across the root to the **xylem vessels** which transport water and mineral salts up to all parts of the plant.

Role of stem

When the cut end of a leafy plant's stem is immersed in red dye (eosin) for an hour and then transverse and longitudinal sections cut, the dye is only found in the xylem vessels showing that xylem (figure 18.3) is the site of water transport in a flowering plant. Since xylem vessels which are hollow and dead, bear rings or spiral bands of strong **lignin**, they also support the plant.

Unlike xylem vessels, the sieve tubes of **phloem** tissue (figure 18.3) are alive. Their cytoplasm is continuous from cell to cell through holes in **sieve plates** thus allowing transport of soluble organic materials. Phloem tissue also contains small **companion cells**, each of which is thought to control the activity of a sieve tube since the latter has no nucleus.

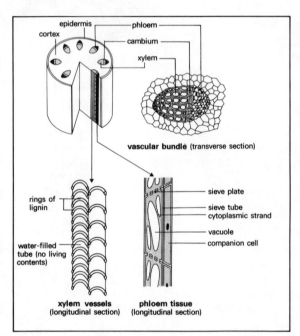

Figure 18.3   Structure of stem (from dicotyledonous plant)

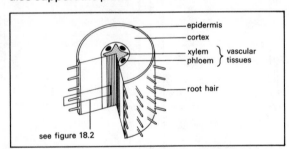

Figure 18.1   Structure of root (from dicotyledonous plant)

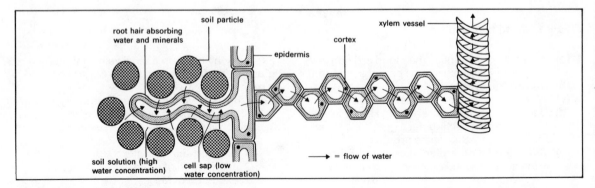

Figure 18.2   Water transport across root

54

Each vascular bundle in a stem also contains **cambium** cells which are able to divide and produce new xylem and phloem tissue. This is especially important in plants which continue to grow for many years and therefore need additional rings of supporting tissue (see chapter 20).

Role of leaf

Transpiration

When the experiment in figure 18.4 is examined after several days, a lot of condensation is found on the inner surface of bell jar A, a little on B and none on C, showing that most water is lost from a plant's leaves. This evaporation of water from a plant is called **transpiration**.

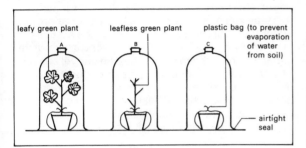

Figure 18.4 Site of transpiration experiment

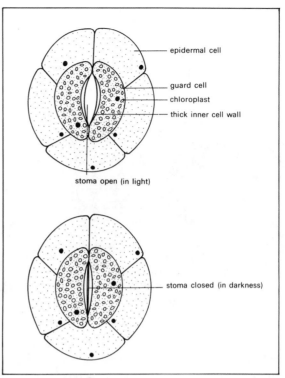

Figure 18.5 Stomata

Stomata

Water vapour escapes from a leaf through tiny pores called **stomata** (figure 18.5). Each **stoma** is bound by two **guard cells** which, unlike the surrounding epidermal cells, are sausage-shaped, contain chloroplasts and have a thick inner cell wall. When the guard cells become **turgid** (e.g. in light) they open the stoma. When they become **flaccid** (e.g. in darkness) they close the slit-like hole.

Location of stomata

Look at the leaves in the experiment shown in figure 18.6. Leaf 1 loses most weight because none of its stomata are clogged with vaseline. Leaf 2 loses no weight because all of its stomata are blocked. Since leaf 3 loses more weight than leaf 4, it is concluded that leaf 3 has fewer blocked stomata and that most stomata must therefore occur on the under surface of a leaf (of a dicotyledonous plant).

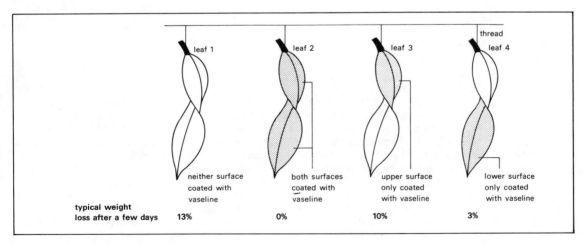

Figure 18.6 Location of stomata experiment

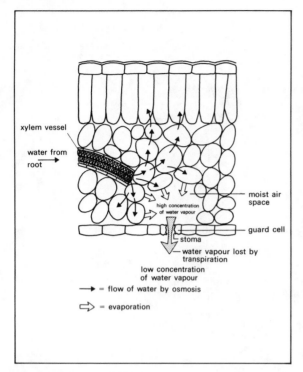

Figure 18.7 Transpiration stream in leaf

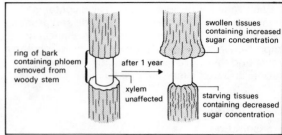

Figure 18.8 Effect of ringing (girdling)

Site of translocation

Ringing involves peeling off a band of bark (which contains the phloem tissue) from a plant such as the young cherry tree in figure 18.8. A plant treated in this way still possesses its xylem (and therefore water transport is unaffected). Since the tissues above the ring are found to swell up and accumulate sugar whereas those below the ring shrivel and die of starvation, it is concluded that **translocation** (transport of soluble organic foodstuffs) normally occurs in phloem tissue but is stopped by ringing. Eventually a ringed plant dies of drought because roots that have died of starvation are no longer able to absorb water.

Under normal circumstances in a plant, sugars are translocated both **downwards** (from leaves to stem and root cells) and **upwards** (from storage organs and leaves to flowers and growing points).

Revision questions

1 Copy and complete the following table:

cell type	description
	divides to form new cells
	thought to be controlled by a companion cell
	supported by a spiral of lignin
	controls the opening and closing of a stoma
	absorbs water from soil solution

This is further verified by submerging a leaf in very hot water. Many small bubbles appear on its underside but only a few appear above. Each bubble indicates the position of a stoma through which hot expanded air has escaped from a moist air space (see figure 18.7).

Since stomata must be in contact with air to allow gaseous exchange, totally submerged plants (e.g. *Elodea*) have no stomata and partially submerged plants (e.g. water lily) have all their stomata on the upper leaf surfaces.

Transpiration stream

This is the name given to the entire route of water through the plant, i.e. the continuous passage of water first entering by the root hairs, then crossing the root cortex, flowing up through the xylem vessels of root, stem and leaf and finally passing by osmosis into leaf cells (figure 18.7). Much water is used to keep the cells turgid, some is used for photosynthesis and the extra gathers in the moist air spaces. From here it evaporates through the stomata by transpiration and keeps the plant cool in hot weather.

In dry weather, however, the amount of water lost by transpiration may exceed that absorbed by the roots. As a result the cells lose their turgid condition causing **wilting** and, over prolonged periods of drought, death of the plant.

2 In which direction, A or B, will the balance shown in figure 18.9 swing? Explain why.

3 When a grass leaf is immersed in very hot water, approximately equal numbers of tiny bubbles appear on both surfaces. What conclusion can be drawn from this experiment?

4 The plant shown in figure 18.10 was kept in bright light for 2 days and then ringed.
  (a) What vascular tissue is undamaged by ringing and what substance does it transport?
  (b) What vascular tissue is removed by ringing and what substance does it normally transport?

(c) After ringing, the plant was left in darkness for 24 hours and then tested for starch. Leaf A gave a positive result, leaf B a negative one. Explain why.

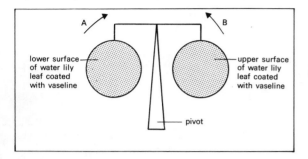

Figure 18.9 see question 2

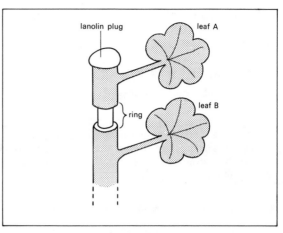

Figure 18.10 see question 4

19 Size and surface area

length of side of cube (cm)	surface area of one side (cm²)	total surface area (cm²) (SA)	volume (cm³) (VOL)	(cm² per SA/VOL cm³)
1	1	6	1	$\frac{6}{1} = 6$
2	4	24	8	$\frac{24}{8} = 3$
3	9	54	27	$\frac{54}{27} = 2$
4	16	96	64	$\frac{96}{64} = 1.5$
5	25	150	125	$\frac{150}{125} = 1.2$
6	36	216	216	$\frac{216}{216} = 1$

Table 19.1 Size and relative surface area

Surface area in relation to size

Measurement of either **volume** (in cm³) or **body mass** (in grams) gives a useful indication of an object's **size**. Its **surface area** is measured in cm². Table 19.1 shows that a small object has a **large** surface area in relation to its size and that a large object has a **small** surface area relative to its size. For example, a one centimetre cube's total surface area is 6 cm² (i.e. 6 cm²/1 cm³) whereas a two centimetre cube's total surface area of 24 cm² is shared by a volume of 8 cm³ (i.e. 24 cm²/8 cm³ or 3 cm²/cm³ on average) and a six centimetre cube's total surface area of 216 cm² is shared by a volume of 216 cm³ (i.e. only 1 cm²/cm³ on average).

Put another way this means that while all 6 cm² of surface area of a one centimetre cube (figure 19.1) are exposed to the outside, only 3 cm²/cm³ of a two

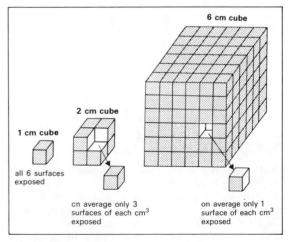

Figure 19.1 Relative surface area

centimetre cube on average are exposed and only 1 cm^2/cm^3 of a six centimetre cube on average is exposed.

Surface area/mass ratio of animals

The approximate surface area of each animal can be determined by measuring the area of a suitable paper cylinder as shown in figure 19.2. When the animals are then weighed and their surface area/mass ratios calculated, the smallest animal is found to have the largest surface area relative to its size and *vice versa.*

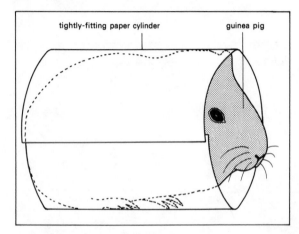

Figure 19.2 Measuring surface area

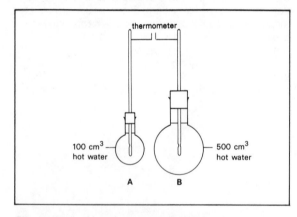

Figure 19.3 Heat loss experiment

Surface area and heat loss

Heat loss from the two flasks shown in figure 19.3 is followed by recording their temperatures at regular intervals and recording the results in a graph. Figure 19.4 shows the **cooling curve** obtained for each flask. The smaller flask is found to cool down more rapidly since it has the greater surface area/volume ratio (see table 19.2).

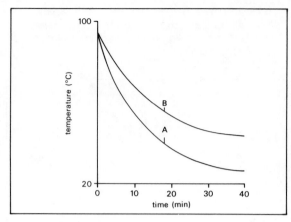

Figure 19.4 Cooling curves

	flask A	flask B
radius (r)	3 cm	5 cm
total surface area ($4\pi r^2$)	$4 \times \frac{22}{7} \times 3^2 = 100$ cm² (approx)	$4 \times \frac{22}{7} \times 5^2 = 300$ cm² (approx)
volume	100 cm³	500 cm³
$\frac{SA}{VOL}$ ratio	$\frac{100}{100} = 1.0$	$\frac{300}{500} = 0.6$

Table 19.2 Relative surface area of 2 flasks

If two animals similar in size to flasks A and B were exposed to extreme cold (and all other factors were constant) then the smaller animal would suffer first because its relatively larger surface area would lose heat more rapidly. However, small mammals consume more food and oxygen relative to their size than larger mammals and their **metabolic** processes (involved in tissue respiration) occur at a faster rate releasing energy to compensate for the constant heat loss. Very few small animals live in extremely cold climates.

Surface area and water loss

Potatoes
When a large and a small potato are peeled and left exposed to the air for a week, the small one is found to lose the greater percentage of its own weight. This is because it has a relatively large surface area/mass ratio from which water is able to evaporate. Similarly a small animal dries out more rapidly than a large one.

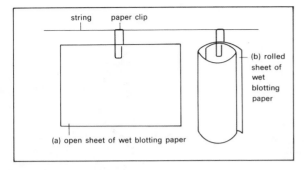

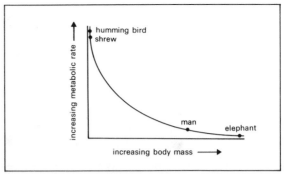

Figure 19.5   Leaf models experiment

Figure 19.7   see question 2

Leaf models

When the 'leaves' in figure 19.5 are blown with a hairdrier, the 'oak leaf' (a) with its large surface area is found to dry out more quickly than the curled 'pine needle' (b). By shedding their leaves in autumn, **deciduous** trees (e.g. oak) prevent such water loss during winter, when the soil water is frozen and cannot be absorbed by the roots.

Revision questions

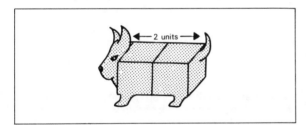

Figure 19.6   see question 1

1 **(a)** Ignoring extremities such as legs, tail, nose etc. calculate the total surface area in 'units' of the young dog shown in figure 19.6.
  (b) Calculate its volume in 'units'.
  (c) What is its surface area/size ratio?
  (d) On reaching old age, the dog was found to have doubled its length, breadth and height. Repeat parts (a)–(c) using these new dimensions.
  (e) At what stage in its life did the dog have the larger surface area in relation to its size?
2 From the graph in figure 19.7 **(a)** describe the relationship that exists between body mass and relative oxygen uptake.
  (b) Which animal has the smallest surface area in relation to its size?
  (c) Explain why a shrew eats about ¾ of its own weight of food per day.
3 Why does it take longer to peel 1 kg of small potatoes than 1 kg of large ones?
4 Suggest why coniferous trees (e.g. Scots pine) do not need to shed all their leaves in autumn.

20 Support in plants and animals

Plants

Compared with a young land (terrestrial) plant (figure 20.1), the stem of a water (aquatic) plant is narrower in diameter since it is supported by the water. Numerous **air spaces** keep the water plant floating near the surface and any tough conducting tissue is found at the stem's centre where it is least likely to break in turbulent water.

A young land plant's stem is supported by tough conducting tissue (xylem, see chapter 18) arranged in a ring of '**reinforcing rods**' and by **turgid** cortex cells pressing against each other. Lack of water may therefore cause such a stem to wilt and collapse.

If a plant continues to grow for many years then a complete **ring** of xylem (rich in lignin) is laid down

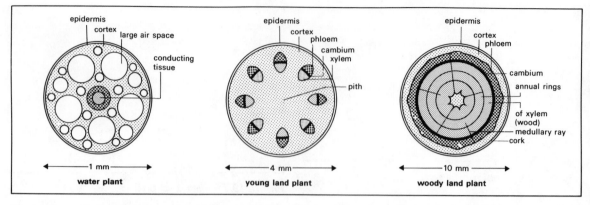

Figure 20.1 Transverse sections of 3 types of stem

every year by division of the **cambium cells**, making the stem gradually become stronger, thicker and very **woody**.

Animals

A **skeleton** is a supporting framework. In many invertebrates it is found on the outside of the body

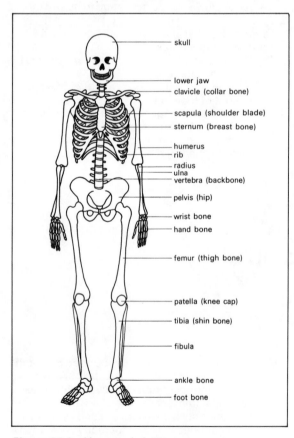

Figure 20.2 Human skeleton

skull
lower jaw
clavicle (collar bone)
scapula (shoulder blade)
sternum (breast bone)
humerus
rib
radius
ulna
vertebra (backbone)
pelvis (hip)
wrist bone
hand bone
femur (thigh bone)
patella (knee cap)
tibia (shin bone)
fibula
ankle bone
foot bone

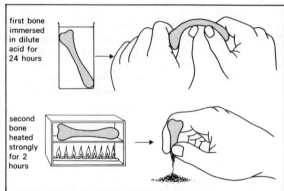

Figure 20.3 Effect of acid and heat on bone

(e.g. shell of an oyster, hard jointed coat of a locust) and is called an **exoskeleton**. In vertebrates it is found inside the body (e.g. bones of man, figure 20.2) and is called an **endoskeleton**. In addition to support, man's skeleton protects vital organs and provides a framework for attachment of muscles thus allowing movement.

Composition of bone
Look at the experiment illustrated in figure 20.3. Bone consists of both **flexible organic** material (which is removed by heat but not by acid) and **hard inorganic** mineral material (mostly calcium phosphate which is removed by acid but not by heat). Since the shape of the bone is unaltered after each treatment, these two components of bone must be closely intermixed.

Although at first glance bone might appear dead, it is very much alive and contains living bone cells (figure 20.4) which build chemicals such as calcium and phosphate (extracted from blood in nearby canals) into hard bony material.

Strength of bone
The **breaking strength** of a 300 mm solid glass rod can be found by performing the experiment shown

in figure 20.5. When a 150 mm length of the same glass rod is used, more weight is required to break it. Similarly a shorter bone is stronger than a longer one.

When a 300 mm length of hollow glass tubing (of equal mass to the original 300 mm solid rod) is used, its breaking strength is also greater than that of the original glass rod. Similarly a hollow (and hence thicker) bone is stronger than a solid (and hence thinner) one of equal mass. Since short thick bones are the strongest, extremely heavy animals (e.g. rhinoceros) are found to have shorter thicker legs than lighter animals (e.g. horse).

and relatively flimsier skeletons than their terrestrial relatives are able to move around and survive in water. If however a blue whale (the world's largest animal) became stranded on land, it would be unable to move and would soon become crushed to death by its own body mass pressing down on its essential organs.

Plants do not have to be able to move from place to place in order to find food. Unlike animals their potential size is not limited by the need for efficient locomotion. The biggest living thing in the world is therefore a plant – the giant redwood tree.

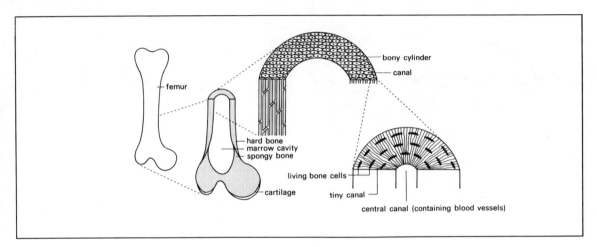

Figure 20.4 Structure of bone

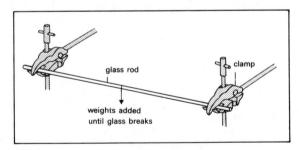

Figure 20.5 Measuring breaking strength

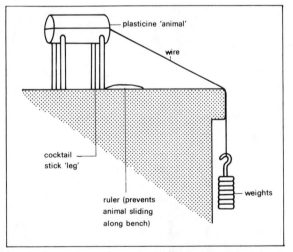

Figure 20.6 Stability experiment

Size limitation

A land animal must support its own body mass. The bigger the animal the shorter and thicker its legs must be in relation to its body mass. However movement on short thick legs is difficult and this puts a limit on the maximum size possible for a land animal.

An aquatic animal's body mass, on the other hand, is supported by the water which exerts an **upthrust** on the animal. Because of this **buoyancy** (tendency to float), mammals with much larger body masses

Stability

An organism is said to be **stable** when it remains in one position and does not fall over. The more stable the organism, the more pulling (or pushing) it can withstand without falling over. When the experiment shown in figure 20.6 is repeated using four shorter legs, more weights are required to make the 'animal' fall over. Shorter legs increase an animal's stability because they bring its **centre of gravity** (point of balance) nearer to the ground.

Figure 20.7 shows the front view of each of three animals. When weights are applied to the corresponding plasticine model of each animal as before, X is found to be the most stable and Z the least stable. This is because X's centre of gravity is

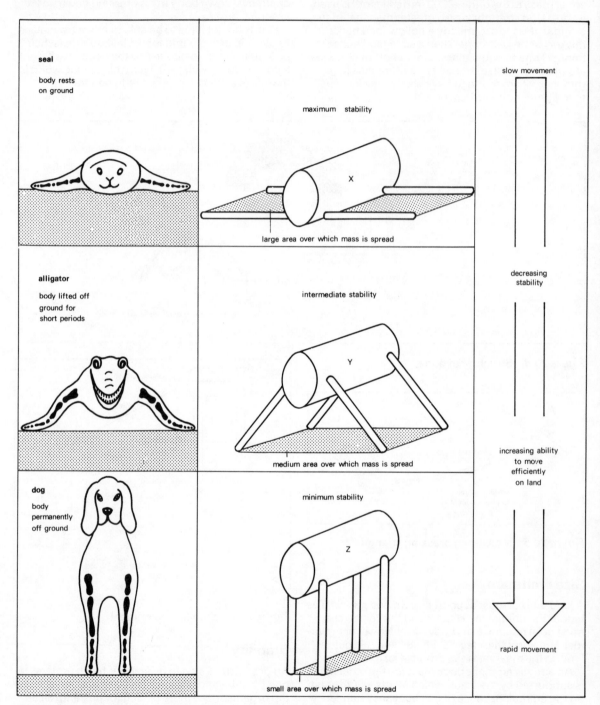

Figure 20.7   Stability versus efficiency of movement on land

nearest to the ground and its mass is spread over the widest area. The reverse is true of Z. However, stability is only one aspect of a creature's success in an environment and animal Z, despite having the poorest stability of the three, is found to be the best adapted to life on land since its limb arrangement lifts its body permanently off the ground and allows rapid movement.

Revision questions

1 Which of the 3 stems shown in figure 20.1 would not collapse during a period of drought? Explain why.
2 Copy and complete the following table:

bone	structure protected
skull	
ribs	
vertebrae	

3 (a) State 2 ways in which the femur of the elephant differs (in relation to the animal's body mass) from the femur of the jack rabbit (hare) shown in figure 20.8.

Figure 20.8 see question 3

(b) What effect does this have on the elephant's rate of movement compared with that of the rabbit?
(c) Why is a femur of this structure essential to an elephant?

4 (a) Consider model Z (figure 20.7). Despite lack of stability, what advantage does a giraffe gain from having its legs arranged the same way?
(b) When a mother giraffe is suckling her young, she stands with her legs splayed out. Give 2 reasons why this increases her stability.

21 Movement in animals

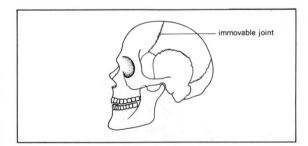

Figure 21.1 Skull (immovable joint)

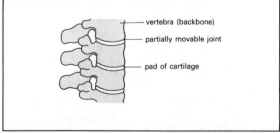

Figure 21.2 Vertebrae (partly movable joint)

Joints

The point of contact between two bones is called a **joint**. There are three main types of joint.

Immovable joints

Since the bones fit tightly together, no movement is possible between them, e.g. **skull** (figure 21.1).

Partially movable joints

Slight movement is possible between the bones, e.g. **vertebrae** (figure 21.2). Pads of **cartilage** prevent friction between the bones and act as shock absorbers.

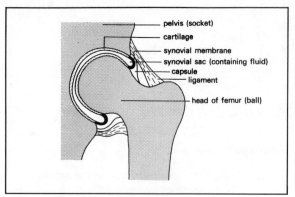

Figure 21.3 Hip (ball and socket joint)

Freely movable joints

Ball and socket

At the **hip** (figure 21.3) and **shoulder**, the rounded head (ball) of one bone fits into the socket of another allowing movement in three planes. Since both ball and socket are lined with smooth slippery cartilage and are separated by a fluid-filled (**synovial**) cavity, friction-free movement is possible. The bones are held together by a **capsule** of tough fibrous tissue and several **ligaments**.

Hinge

At the **knee** and **elbow** (figure 21.4), the bones meet as a hinge and movement is restricted to one plane.

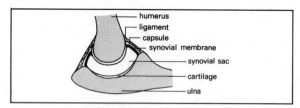

Figure 21.4 Elbow (hinge joint)

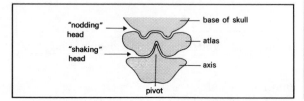

Figure 21.5 Rocking and pivotting joints

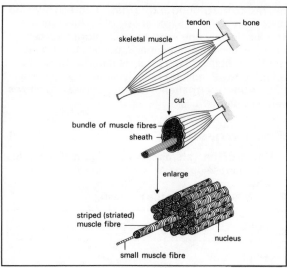

Figure 21.6 Skeletal muscle

Other types

The wrist bones move by **gliding** across each other. Nodding movements of the head occur by the skull **rocking** to and fro on the first vertebra, the atlas (figure 21.5). Sideways movements (shaking) of the head occur by the atlas **pivotting** on the second vertebra, the axis.

Muscles

The three types found in the body are **skeletal**, **smooth** and **heart** muscle.

Skeletal muscle

A bundle of skeletal muscle **fibres** (figure 21.6) is attached at each end by tendons to the skeleton. It contracts rapidly in response to stimulation by nerves and pulls on bones producing movement. Since it is controlled by the voluntary part of the nervous system, it is called **voluntary** muscle. Compared with other muscle types, it fatigues quickly.

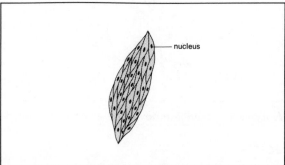

Figure 21.7 Sheet of smooth muscle cells

Smooth muscle

This consists of sheets of **spindle-shaped cells** (figure 21.7) which are found in the walls of the body's muscular tubes (e.g. arteries and intestines). It contracts slowly bringing about opening and closing movements such as peristalsis (see chapter 8). Since it is controlled by the involuntary part of the nervous system, it is called **involuntary** muscle. It fatigues slowly.

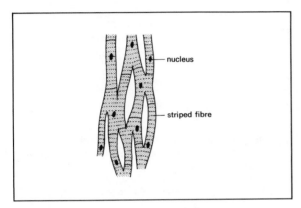

Figure 21.8 Heart (cardiac) muscle

Heart (cardiac) muscle

Although consisting of striped (striated) fibres (figure 21.8), heart muscle is **involuntary**. Its rhythmic contractions are generated within itself and they normally occur about 70 times per minute. It never becomes tired.

Movement of a vertebrate limb

Figure 21.9 shows the arrangement of the pair of skeletal muscles that operate a hinge joint. Contraction of the **flexor** muscle bends the limb whereas contraction of the **extensor** straightens it. Two such muscles, which produce movement of the limb in opposite directions, make up an **antagonistic pair**. A muscle is able to contract (and then later become relaxed) but is unable to actively lengthen of its own accord. Thus, following contraction, a muscle depends on the action of its antagonistic partner to restore it to its original length.

Movement of an invertebrate limb

Whereas in a vertebrate the two antagonistic muscles are attached by tendons to the outside of a bony **endoskeleton**, in an invertebrate (figure 21.9) they are attached by internal processes to the inside of an **exoskeleton** (made of chitin in arthropods). Although an invertebrate has 2 peg and socket joints instead of a hinge joint, muscular contraction brings about similar movement of the limb in one plane.

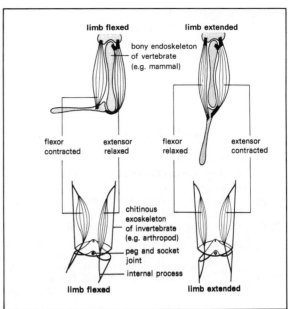

Figure 21.9 Movement of vertebrate and invertebrate limbs

Posture

When the body is at rest, the muscles remain in a state of **semi-tension** and hold the body in a certain posture. Good posture (standing or sitting up straight) should always be adopted since it allows easy breathing and tires the muscles much less than when the body is slouched.

Revision questions

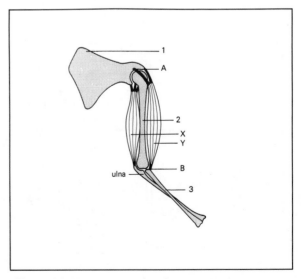

Figure 21.10 see question 1

65

1 Look at figure 21.10.
   (a) Name bones 1, 2 and 3.
   (b) Using all of the following words and phrases, write 1–2 sentences to describe the difference between joint types A and B:
   ball; hinge; joint; movement; one plane; socket; three planes.
   (c) Why is a shoulder more easily dislocated than a hip?
   (d) Name muscles X and Y.
   (e) What effect will contraction of muscle X have on this limb?
   (f) When the arm is fully bent which muscle is relaxed?
   (g) Explain why the muscles in the diagram are said to be in a state of semi-tension.
2 Copy and complete the following table:

muscle type	fibrous or non fibrous	fast or slow to fatigue	one place where found in body
skeletal			
smooth			
cardiac			

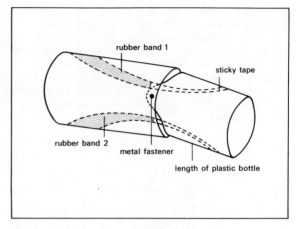

Figure 21.11   see question 3

3 Figure 21.11 shows a model of a limb.
   (a) To which of the following animals would such a limb belong?
   Tortoise, locust, monkey, earthworm.
   (b) Match each label in the diagram with one of the following:
   exoskeleton; extensor; flexor; internal process; peg and socket joint.

22 Movement in plants

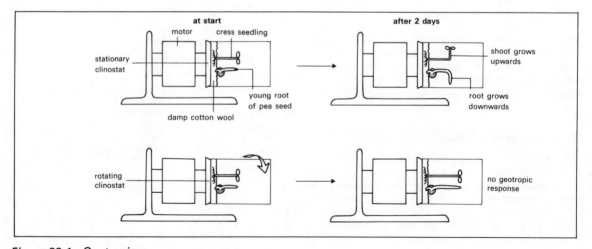

Figure 22.1   Geotropism

Tropism

A tropism is a directional growth movement of a plant organ in response to a stimulus from one direction.

Geotropism

The experiment in figure 22.1 shows the effect of the stimulus **gravity** on the growth of a root and a shoot. In the stationary clinostat, roots grow down towards the stimulus source. This response is called **positive geotropism**. On the other hand, shoots exhibit **negative geotropism** by growing upwards away from the stimulus. In the rotating clinostat, the control experiment, gravity acts equally on all sides of each plant organ and therefore no geotropic growth movements occur.

Phototropism

The experiment in figure 22.2 shows that shoots exposed to **unilateral light** bend towards it. This response is called **positive phototropism**. The shoots on the rotating clinostat receive even illumination on all sides and therefore show no phototropic movements. The third part of this experiment demonstrates that it is the tip of a shoot that detects the light.

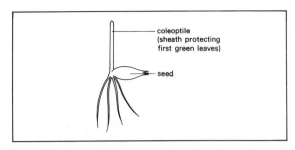

Figure 22.3   Oat seedling

Plant hormone (growth substance)

Oat coleoptiles (figure 22.3) are used in the experiments shown in figure 22.4 from which the following conclusions are made. The shoot tip is essential for growth and produces a **chemical**. This chemical messenger diffuses downwards to lower regions of the shoot where it stimulates growth by making the cells elongate. This growth substance, which can diffuse through agar (and gelatin) but not through metal, is now known to be a plant hormone (**auxin**). The most common auxin is called **indole acetic acid (IAA)**. When IAA is applied to one side of a coleptile (figure 22.5), that side grows at a faster rate than the untreated side causing the shoot to bend.

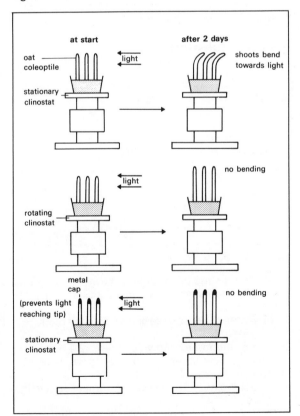

Figure 22.2   Phototropism

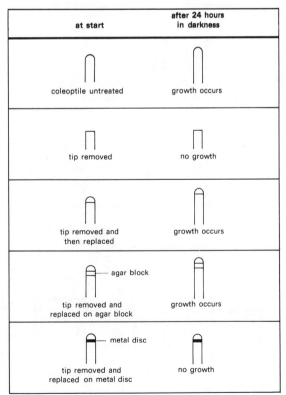

Figure 22.4   Coleoptile experiments

67

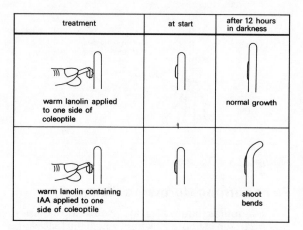

treatment	at start	after 12 hours in darkness
warm lanolin applied to one side of coleoptile		normal growth
warm lanolin containing IAA applied to one side of coleoptile		shoot bends

Figure 22.5   Effect of IAA on coleoptile growth

Mechanism of phototropism

When a shoot is exposed to unilateral light (figure 22.6), an unequal distribution of hormone (auxin) occurs in the region of the shoot just behind the tip. More auxin gathers in the non-illuminated side and therefore more growth occurs on that side, making the shoot bend towards the light.

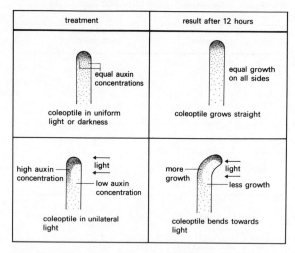

treatment	result after 12 hours
coleoptile in uniform light or darkness — equal auxin concentrations	equal growth on all sides — coleoptile grows straight
coleoptile in unilateral light — high auxin concentration, low auxin concentration	more growth, less growth — coleoptile bends towards light

Figure 22.6   Hormonal explanation of phototropism

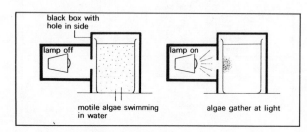

Figure 22.7   Positive phototaxis

Taxis

A taxis is a movement of an organism directly towards a stimulus from one direction. Motile algae exhibit **positive phototaxis** by moving directly towards light (figure 22.7).

Nastic movements

A nastic movement is a non-directional movement of plant parts in response to an external stimulus (e.g. changé in temperature) which acts equally all round the plant (see figure 22.8).

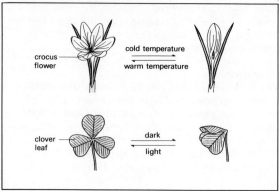

Figure 22.8   Nastic movements

Revision questions

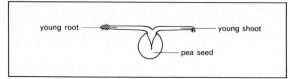

Figure 22.9   see question 1(a)

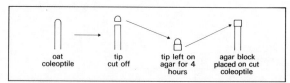

Figure 22.10   see question 3

1 (a) Redraw the pea seedling shown in figure 22.9 after two days growth in a dark stationary place.
   (b) Name the response shown by each plant organ.
2 If the 5 coleoptiles shown in figure 22.4 were all illuminated unilaterally which ones would bend towards the light?
3 Predict the outcome of the experiment shown in figure 22.10.
4 The prefix *hydro* means water. Which plant organ exhibits positive hydrotropism?

23 Sense organs

Sensitivity is the ability of an organism to respond when acted upon by a **stimulus**. In man the sense organs contain **receptors** which pick up environmental stimuli. These are passed as impulses to the nervous system (chapter 24).

Skin

The five types of receptor contained in the skin are shown in figure 23.1. These receptors are not spread evenly throughout the body. For example some areas have more touch receptors than others and therefore allow more accurate discrimination between, say, the two points of a pair of dividers held 1 mm apart, than do other regions of the skin.

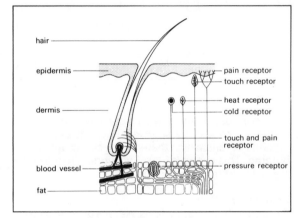

Figure 23.1 Skin receptors

part	structure and function
conjunctiva	Thin transparent membrane. Covers front of eye and inside of eyelids.
sclerotic	'White' of eye. Tough protective layer.
cornea	Transparent continuation of sclerotic at front of eye. Allows light to enter.
choroid	Black layer containing blood capillaries. Supplies food and oxygen to eye and reduces reflection of light inside eyeball.
iris	Coloured continuation of choroid. Contains muscles which alter diameter of pupil.
pupil	Opening surrounded by iris. Controls amount of light entering eye. Activity of iris muscles enlarge it in dim light to admit maximum light and reduce it in bright light to admit minimum light thus preventing damage to retina.
lens	Flexible transparent biconvex structure. Focuses light on to retina.
ciliary muscle	Ring of muscle fibres surrounding lens. Alters curvature of lens allowing near and distant objects to be focused on to retina (see figure 23.3).
aqueous humour	Transparent liquid. Helps to focus light and maintain shape of eye.
vitreous humour	Transparent jelly. Maintains shape of eye.
retina	Light-sensitive layer containing 2 types of receptor cell: **rods** (sensitive to dim light) and **cones** (sensitive to bright light and colour). Light converted to nerve impulses.
fovea	Small depression containing cones. Point of most accurate vision.
blind spot	Point where receptors are absent and fibres from retina enter optic nerve.
optic nerve	Carries nerve impulses from retina to brain.
eye muscles	Six muscles attached to sclerotic and bony eye socket. Allow movement of eyeball in socket.

Table 23.1 Functions of parts of eye

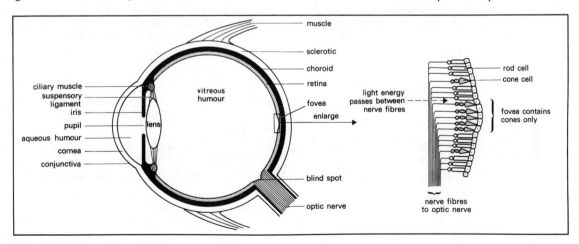
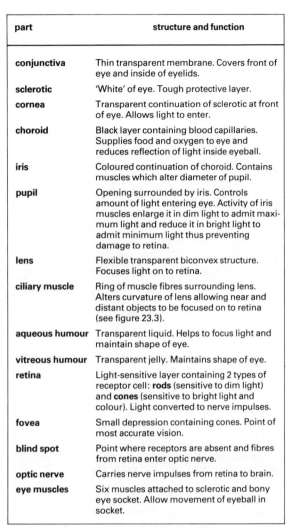

Figure 23.2 Structure of eye

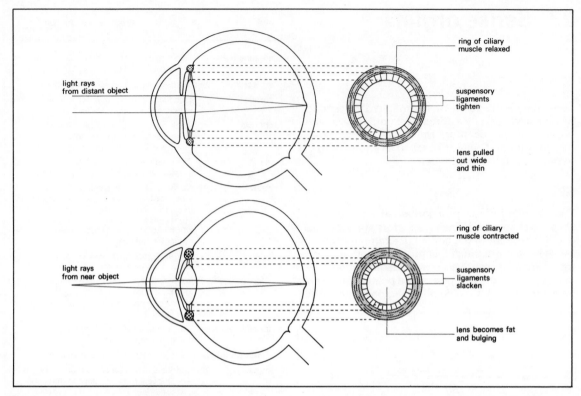

Figure 23.3 Accommodation of lens

Eye

Table 23.1 gives a summary of the functions of the parts of the eye shown in figure 23.2.

Accommodation (adjustment) of lens
When the ring of **ciliary muscle** (figure 23.3) becomes relaxed, the diameter of the circle it forms increases. As a result the **suspensory ligaments** become taut and pull the lens out into a **wide thin** shape which is ideal for bringing to a focus rays of light from a distant object.

On contraction, the ring of ciliary muscle decreases in diameter and the suspensory ligaments no longer pull on the lens which therefore becomes **fat** and **bulging**. This shape is ideal for bringing to a focus rays of light from a near object.

Image formation
Light travels in straight lines. As rays of light from an object enter the eye, they are bent (refracted) by the cornea and lens and brought to a focus on the retina as an **inverted** image (figure 23.4). The inversion of the image is later corrected by the brain.

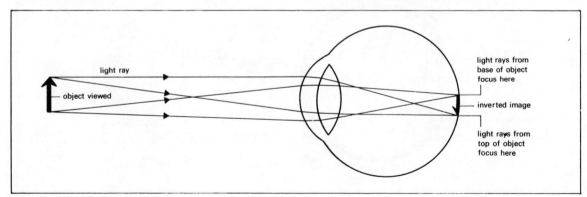

Figure 23.4 Image formation

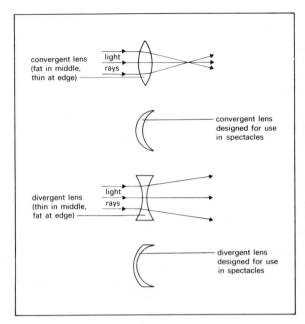

Figure 23.5   Types of lens

Correction of long and short sight

Whey rays of light pass through a **convergent** lens (figure 23.5) they tend to come together; the reverse is true of a **divergent** lens. Long sight is caused by the eyeball being shorter than normal. The point of focus for a near object occurs behind the retina (figure 23.6). It is corrected by a convergent lens. Short sight is caused by the eyeball being longer than normal. The point of focus for a distant object occurs in front of the retina. It is corrected by a divergent lens.

Ear

Hearing

Table 23.2 describes the parts of the ear involved in hearing as illustrated in figure 23.7.

Balance

Located in the inner ear are the three liquid-filled **semi-circular canals** (figure 23.7) situated at right angles to one another. Each possesses an **ampulla** containing hair-like receptors which are stimulated

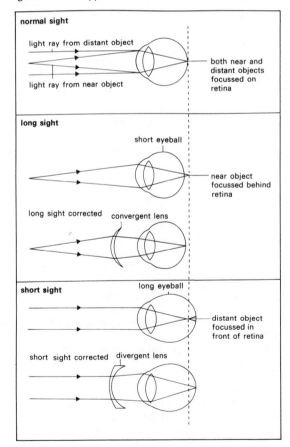

Figure 23.6   Correction of long and short sight

part	structure and function
pinna	Flap of skin and cartilage. Collects sound waves.
auditory canal	Air-filled tube. Directs sound waves on to eardrum.
eardrum	Thin membrane stretched completely across end of auditory canal. Set vibrating by sound waves which it passes on to middle ear bones.
hammer, anvil and **stirrup**	Three tiny bones. Amplify and transmit sound vibrations from eardrum to oval window.
Eustachian tube	Tube connecting middle ear to throat. Maintains equal air pressure on either side of eardrum. This prevents eardrum from bursting and allows it to vibrate properly.
oval window	Thin membrane. Transmits sound vibrations into liquid-filled inner ear.
cochlea	Coiled liquid-filled tube lined with sound **receptor** cells possessing hair-like endings. Stimulation of latter by sound vibrations in the liquid leads to nerve impulses being sent via the **auditory nerve** to the brain.

Table 23.2   Functions of parts of ear involved in hearing

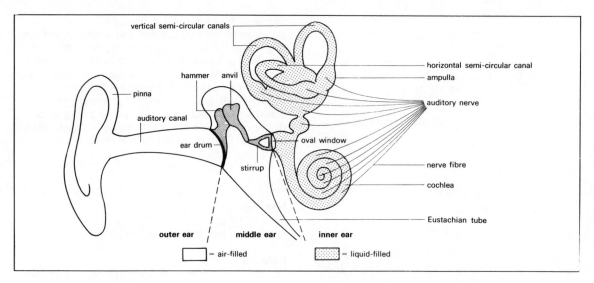

Figure 23.7 Structure of ear

when the ampulla and its canal are rotated in their respective plane (see figure 23.8). Messages are sent via the auditory nerve to the brain which controls the muscular activity essential for balance and posture. For accurate sensations of balance human beings also rely on information from their eyes and joints.

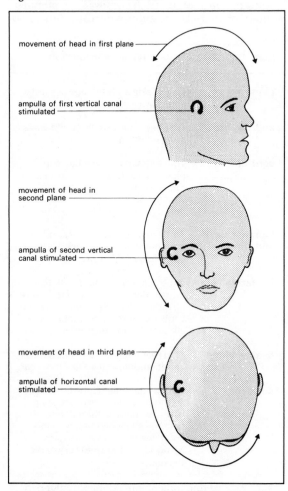

Figure 23.8 Role of semi-circular canals

stimulus (or agent causing stimulus)	sense organ	receptor	sense
light	eye	light receptor cells in retina (rods and cones)	sight
sound	ear	sound receptor cells in cochlea	hearing
rotation of head	ear	movement receptor cells in ampullae of semi-circular canals	balance
chemical liquid	tongue	chemical receptor cells in taste buds	taste (sweet, sour, bitter, salty)
chemical gas	nose	chemical receptor cells in nasal lining	smell
physical contact	skin	touch receptors	touch
low temperature	skin	cold receptors	cold
high temperature	skin	heat receptors	heat
mechanical stress	skin	pressure receptors	pressure
injury	skin	pain receptors	pain

Table 23.3 Summary of senses

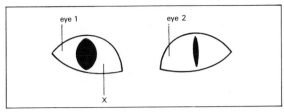

Figure 23.9   see question 2

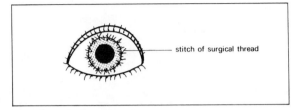
stitch of surgical thread

Figure 23.10   see question 3

Revision questions

1 (a) Name man's five senses as described traditionally.

(b) Briefly explain why, strictly speaking, it is inaccurate to say that man has five senses.

2 Figure 23.9 shows two eyes of a cat. Which is adapted to (a) bright (b) dim light? (c) Name tissue X.

3 Name the tissue that has been grafted on to the human eye in figure 23.10.

4 (a) List the following in the correct order in which a ray of light on entering the eye would pass through them and then opposite each state its function:
   aqueous humour; conjunctiva; pupil; vitreous humour; cornea; lens.

(b) List the following in the correct order in which a sound vibration on entering the ear would pass through them:
   anvil; auditory canal; eardrum; stirrup; cochlea; oval window; hammer.

5 Prior to modern medical surgery which of the following were (a) temporary (b) permanent forms of deafness? Absence of anvil; cattargh in Eustachian tube; unequal air pressure on either side of eardrum; closed oval window.

24 Nervous system

Man's central nervous system (CNS) is connected to all parts of the body by nerves (figure 24.1) and brings about co-ordination between the body's **receptors** and **effectors** (figure 24.2). This ensures that all of the body's organs and systems work together as an integrated whole.

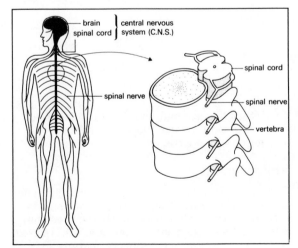

Figure 24.1   Human nervous system

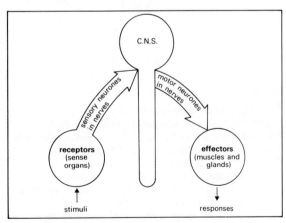

Figure 24.2   Role of nervous system

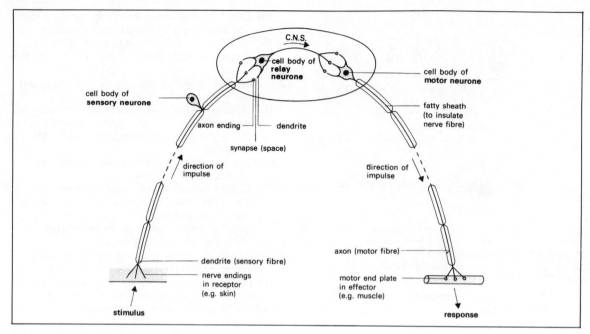

Figure 24.3   The reflex arc

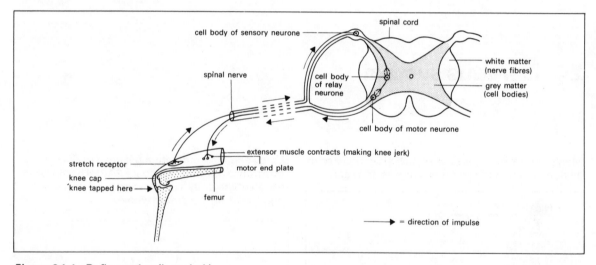

Figure 24.4   Reflex action (knee jerk)

Neurones

The nervous system is made of nerve cells (**neurones**). The simple arrangement of three neurones shown in figure 24.3 is called a **reflex arc**. Nerve impulses are carried to cell bodies by **dendrites** and away from them by **axons**. A tiny space (**synapse**) occurs between the axon ending of one neurone and the dendrite of the next. When a nerve impulse arrives, the knob at the end of the axon branch releases a **chemical** which diffuses across the space and triggers off an impulse in the

dendrite of the next neurone in the arc. Pain-killer drugs (e.g. morphine) produce their effect by interfering with the chemical.

Reflex action

The transmission of a nerve impulse through a reflex arc results in a **reflex action**. In figure 24.4 stretch receptors respond to the tapping of the knee (the **stimulus**) by sending out an impulse which is picked up by the dendrites of the sensory neurone and passed via its cell body to its axon. The impulse then crosses the first synapse and is picked up by the dendrites of the relay neurone (in the spinal cord) which passes it on into its axon. Having crossed the second synapse the impulse is finally picked up by the motor neurone which quickly conducts it down its axon (located in a spinal nerve) to the motor end plates in the muscle where a chemical is released bringing about muscular contraction (the response). This makes the leg kick out (i.e. knee jerk).

A reflex action is a rapid, automatic, involuntary response to a stimulus. It does not require conscious thought by the brain. Thus many reflex actions may still be performed for a short period by an animal whose brain has been destroyed. Reflex actions are protective in function (see tables 24.1 and 2).

reflex action	stimulus	response	protective function
blinking	object touching eye surface	contraction of eyelid muscle	prevents damage to eye
withdrawal of limb (e.g. arm)	pain (e.g. intense heat)	contraction of flexor muscle	prevents damage to limb
sneezing	foreign particles in nasal tract	sudden contraction of ;hest muscles	removes unwanted particles from nose

Table 24.1   Reflex actions whose responses can be partly altered by voluntary means

reflex action	stimulus	response	protective function
peristalsis	presence of food in gut	muscular contraction of gut wall	ensures movement and therefore efficient digestion of food
dilation of eye pupil	dim light	movement of iris muscle	improves vision in poor lighting
paling of skin ('blue with cold')	drop in body temperature	contraction of skin capillaries	reduces heat loss since blood diverted away from skin

Table 24.2   Reflex actions whose responses are completely involuntary

Brain

The brain consists of several different regions (figure 24.5). The **medulla** controls the rate of breathing and heartbeat. The **cerebellum** controls balance and muscular co-ordination. The left and right cerebral hemispheres make up the **cerebrum** which contains about 90% of the body's neurones in its folded outer layer of grey matter.

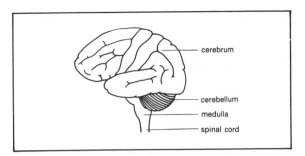

Figure 24.5   Human brain

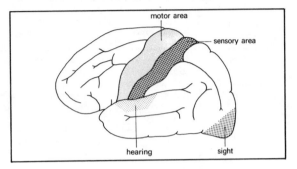

Figure 24.6   Left cerebral hemisphere

Each area of the cerebrum (figure 24.6) is concerned with a specific function. The **sensory area**, for example, receives impulses from the sense organs. In the light of this information, the **motor area** transmits impulses to all parts of the body making them work together efficiently. The unlabelled areas in figure 24.6 are concerned with mental processes such as memory, imagination, reason, conscious thought and intelligence.

Revision questions

1 Rearrange the six terms given below to indicate the correct route taken by a nervous impulse passing from the axon of a sensory neurone to the dendrite of a motor neurone in a reflex arc. dendrite; synapse; axon; synapse; cell body of relay neurone

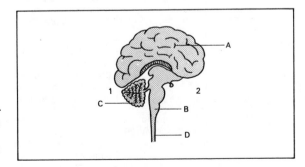

Figure 24.7   see question 3

reflex responses under partial voluntary control	involuntary reflex responses

Copy and complete the above table using the following examples.
coughing; contraction of eye pupil; flushing of skin; laughing; churning of food in stomach; muscular contraction making knee jerk.

3 Figure 24.7 shows a section of a human brain.
 (a) Would the person's face be at side 1 or 2?
 (b) Name parts A, B, C and D and match them with the following descriptions.
   controls pulse rate
   co-ordinates movement during skating
   relays impulses causing spinal reflexes
   responsible for understanding this question

25 Animal hormones

endocrine gland	hormone	effect	notes
pituitary	growth hormone	stimulates growth of developing animal	over-secretion results in giantism; under-secretion leads to dwarfism
pituitary	thyroid-stimulating hormone	controls thyroid gland and stimulates thyroxin production	one of many pituitary hormones which regulate the other endocrine glands
thyroid	thyroxin	(a) controls rate of growth and development (b) controls rate of metabolism	over-secretion in adults causes over-activity and thinness; under-secretion in adults leads to sluggishness and overweight
pituitary	follicle-stimulating hormone (FSH)	(a) stimulates development of follicle in ovary (b) stimulates ovary tissue to produce oestrogen	Pieced together these events make up the menstrual cycle (figure 25.2) which involves the regular alternation of ovulation (egg release) and menstruation (breakdown and release of uterus lining and unfertilised egg). If fertilisation occurs, the corpus luteum persists and continues to secrete progesterone which sustains the uterus wall during early pregnancy.
ovary	oestrogen	(a) heals and repairs uterus (b) stimulates pituitary to make luteinising hormone	
pituitary	luteinising hormone (LH)	(a) brings about ovulation (b) causes follicle to become corpus luteum which secretes progesterone	
ovary (corpus luteum)	progesterone	promotes growth of uterus wall	
testis	testosterone	controls growth and development of male sex organs and secondary sexual characteristics	testis is in turn controlled by a pituitary hormone (LH)
islets of Langerhans in pancreas	insulin	controls conversion of excess glucose in bloodstream to glycogen (stored in liver)	deficiency leads to *diabetes mellitus* (glucose not stored as glycogen but instead lost in urine)
adrenal	adrenaline	(a) increases rate of heartbeat and breathing (b) diverts blood to skeletal muscles (c) causes conversion of stored glycogen to glucose	prepares the body for energy-demanding 'fight' or 'flight' during an emergency
pituitary	anti-diuretic hormone (ADH)	brings about osmo-regulation by controlling amount of water re-absorbed into blood from kidney tubules	see figure 25.3

Table 25.1 Hormones and their effects

In addition to the control effected by the nervous system (chapter 24), further co-ordination of the workings of the human body is brought about by **hormones**. These chemical messengers are secreted directly into the bloodstream by **endocrine** (ductless) glands (figure 25.1). When a hormone reaches a particular part of the body (normally a target organ) it evokes a specific response as summarised in table 25.1.

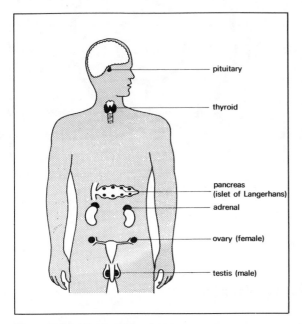

Figure 25.1 Endocrine glands

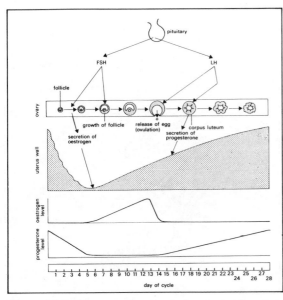

Figure 25.2 Hormonal control of menstrual cycle

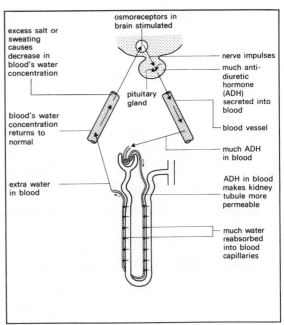

Figure 25.3 Osmoregulation in man

Nervous and hormonal co-ordination differ in that the nervous system (with its speedy transmission of impulses along nerves) enables the body to make rapid, short-lived responses (e.g. reflex actions) whereas the endocrine system (with its dependence on blood for hormone transport) controls slower, longer-lived responses (e.g. those involved in the menstrual cycle).

Osmoregulation and homeostasis

When lack of drinking water, profuse sweating or intake of excess salt causes a decrease in the blood's water concentration, the sequence of events shown in figure 25.3 occurs, very concentrated urine is produced and the blood's water concentration returns to normal.

When excessive intake of water causes an increase in the blood's water concentration, less ADH is made, less water is reabsorbed, dilute urine is produced and again the blood's normal water concentration is restored.

Such maintenance of constant conditions by a receptor responding to a change and then feeding back a message (in this case via the hormone ADH) to an effector whose response brings about a return to normal conditions is called **feedback control** or **homeostasis**. Osmoregulation is only one of many examples of homeostasis by which the body maintains its internal environment at a constant optimum state (regardless of changes in the external environment).

Revision questions

1 One of the endocrine glands regulates the activity of many of the other endocrine glands. Name this so-called 'master' gland.
2 Despite adequate sugar in the diet, the blood sugar level of a person suffering from *diabetes mellitus* can drop to 40 mg/100 cm³.
  (a) By what means does the sugar leave the body?
  (b) What hormone is in deficient supply?
  (c) What chemical conversion does this hormone normally cause the liver to bring about?
3 **(a)** A gonadotrophic hormone stimulates gonads (reproductive organs). Name two such hormones found in human females.
  (b) An ovarian hormone is made by ovary tissue. Name two such hormones.
  (c) Match the above four hormones with black arrows 1, 2, 3 and 4 in figure 25.4.
  (d) Which of these hormones stimulates ovulation?
4 An exocrine gland is a gland which does have ducts. Explain why the pancreas can be regarded as both an endocrine and exocrine gland.

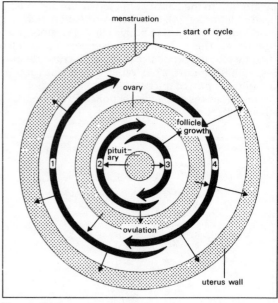

Figure 25.4   see question 3

26 Sexual reproduction in animals

Reproduction is the production of new members of a species. During sexual reproduction fertilisation occurs by two sex cells (**gametes**) and their nuclei fusing to form a **zygote** which develops into a new organism. During asexual reproduction (chapter 30) offspring are derived from a single parent without fertilisation or gamete production.

Gametes
A **sperm** is a tiny male sex cell (gamete). It consists of a head region (mainly a nucleus containing genetic material) and a tail which enables it to move.
   An **ovum** (egg cell) is a female gamete. It cannot move of its own accord and is larger than a sperm because, in addition to its nucleus, it contains a store of food in its cytoplasm.

External fertilisation

Keelworm (*Pomatoceros*)
These worms live inside protective tubes (figure 26.1) cemented to rocks in the sea and the two sexes

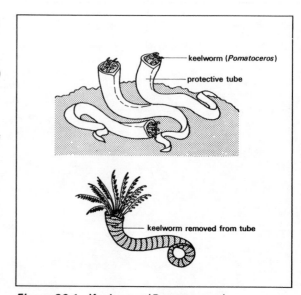

Figure 26.1   Keelworm (*Pomatoceros*)

do not meet. Instead the worms shed their gametes directly into the sea water. Most gametes are lost however and the chance of fertilisation is low. To ensure that at least a few eggs are fertilised, thousands of eggs and millions of sperm have to be released. This is therefore both a wasteful method and one which offers no protection to the fertilised eggs from predators.

Herring
At certain times of the year male and female herring gather in huge numbers near the sea bed. Thousands of eggs are laid and, as they sink to the bottom, millions of sperm are released over them by the males. Many sperm find an egg to fertilise. Thus the chance of fertilisation is increased by the establishment of a **breeding season** which brings the two sexes together and reduces the distance to be travelled by the sperm.

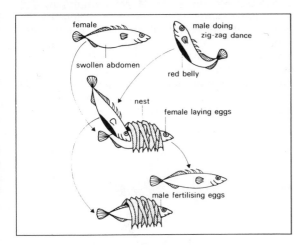

Figure 26.2   Mating in sticklebacks

Stickleback
During the breeding season the male builds a **nest** of waterweeds and develops a **red belly** which attracts a female. The two fish perform **courting** movements (e.g. zig-zag dance, figure 26.2) and finally the female lays her eggs in the nest. The male then sheds his sperm into the nest fertilising many of the eggs. Thus the chance of fertilisation is further increased by **courtship behaviour** of the parents and the nest as a meeting place for the gametes.

Frog
During the breeding season the male attracts the female by 'singing' to her and then clings to her back (figure 26.3) using claspers. As soon as the female releases her eggs into the water, the male sheds his sperms over them. Thus the chance of fertilisation is further increased by the **physical proximity** of the parents.

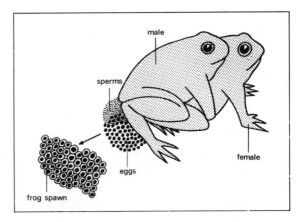

Figure 26.3   Mating in frogs

Internal fertilisation

In terrestrial (land-living) animals ova and sperm are not released into water. Instead a fluid containing sperm is produced by the male and transferred directly into the female's body during **copulation**.

Housefly
Mating involves the ends of the two animals' abdomens coming together (figure 26.4) and sperm passing from male to female. This makes the chance of fertilisation very high.

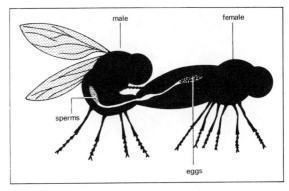

Figure 26.4   Copulation in houseflies

Bird
Following courtship and nest building, the male mounts the female. This copulation (figure 26.5) brings the birds' reproductive openings together and sperm pass into the female. Internal fertilisation occurs and each egg receives a shell before being laid. Thus the chance of fertilisation is very high and the fluid-filled eggs are protected from rapid desiccation (drying out) by their shells and to some extent from predators by the parents.

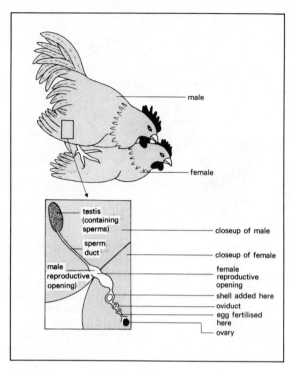

Figure 26.5   Copulation in fowl

Mammal

Human reproductive organs are shown in figure 26.6. Ovulation (release of eggs from the ovary) and the changes that accompany it in a female mammal's body make up the **oestrous** cycle. In some mammals (e.g. dogs) this only occurs during the breeding season whereas in others (e.g. rats and humans) it occurs all the year round.

At around the time of ovulation a woman's uterus wall develops a spongy lining which is rich in blood vessels and ready to receive an egg should one be fertilised. If fertilisation does not occur, this special lining breaks away along with a little blood and passes out through the vagina. This process is called **menstruation** (see also chapter 25).

A male mammal has a special organ, the **penis**, for depositing sperm into the female. During copulation (sexual intercourse in humans) the penis (stiff and erect since it has received an extra supply of blood) is inserted into the vagina. Muscles around the testes and sperm ducts finally force millions of sperm up the ducts and out into the upper end of the vagina near the uterus. The sperm swim up the uterus and into the oviducts. It is here that one sperm fertilises one egg (figure 26.7) forming a zygote. The zygote becomes an embryo by undergoing rapid cell division (**cleavage**) while it is moving towards the uterus into which it soon becomes embedded (im-

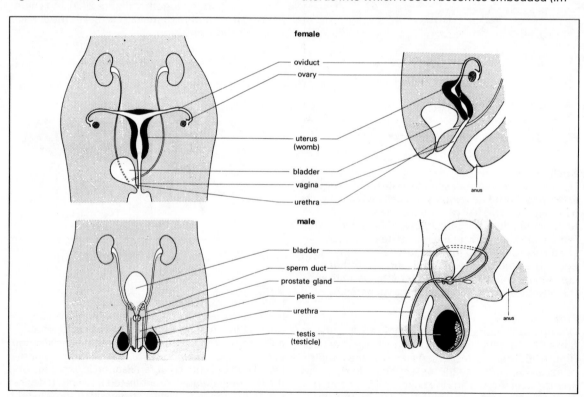

Figure 26.6   Human reproductive organs

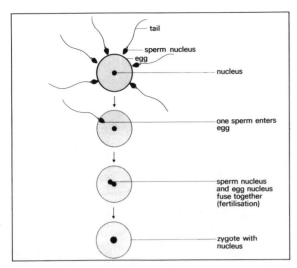

Figure 26.7   Fertilisation

planted) and begins its period of **gestation** (see chapter 27).

Thus in mammals the chance of fertilisation is very high indeed and since the fertilised eggs are normally retained inside the female's body, they are protected from desiccation, extremes of temperature and predators.

external fertilisation	internal fertilisation
found amongst aquatic (water) animals since water available for sperm to swim to eggs	found amongst terrestrial (land) animals since no water present for sperm to swim to egg
many gametes made because relatively inefficient and wasteful method	fewer gametes made because more efficient and less wasteful method
fertilised eggs receive little or no protection	fertilised eggs receive protection from desiccation (and often from predators) by being surrounded by moist sand (locust), a waterproof shell (chick) or the mother's body (man)

Table 26.1   Comparison of external and internal fertilisation

Revision questions

1 (a) Arrange the following animals in order of increased chance of fertilisation:
    toad, polar bear, trout.
   (b) In which of these is fertilisation external?
   (c) In which of these does the egg receive most protection?
   (d) Which of these animals releases the most eggs at a time?
2 Compare a keelworm and a housefly with respect to (a) type and (b) efficiency of method of fertilisation.
3 With reference to figure 26.8, copy and complete the following table:

site of	number of structure	name of structure
copulation		
egg production		
fertilisation		
embryo development		

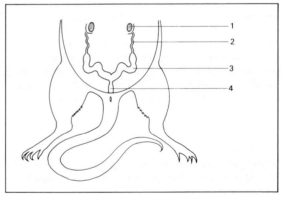

Figure 26.8   see question 3

27 Growth and development of animals

A species' **life cycle** consists of the complete sequence of stages that occur from the zygote of one generation to the same stage of the next generation.

Metamorphosis is the name given to the series of stages by which an immature animal changes into an adult.

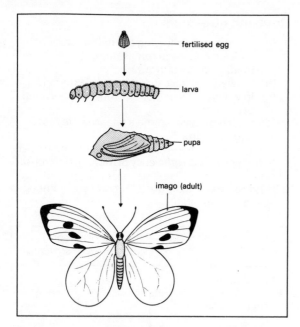

Figure 27.1 Complete metamorphosis (cabbage white butterfly)

Insects

The first type of insect life cycle always involves **complete metamorphosis** (figure 27.1). When the fertilised egg hatches, the **larva** which emerges and the **pupa**, into which it later develops, are both entirely different from the **adult**. The larva eats and moults and finally, when fully grown, turns into the pupa. Many insects survive winter as a motionless pupa. During this time, internal reorganisation occurs and the sexually mature adult (**imago**) is formed. It emerges in spring to repeat the cycle.

The second type of insect life cycle involves **incomplete metamorphosis** (figure 27.2). When the fertilised egg hatches, the **nymph** that emerges

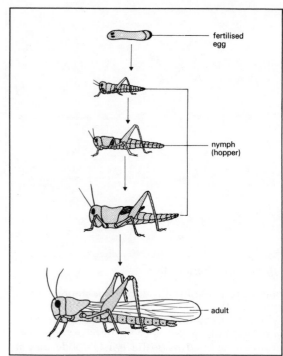

Figure 27.2 Incomplete metamorphosis (locust)

closely resembles the adult except that it is smaller and lacks wings and mature reproductive organs. Again the young insect eats vast amounts of food,

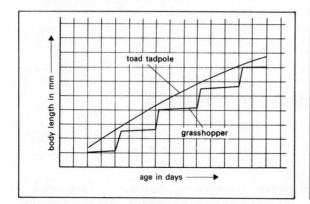

Figure 27.3 Types of growth curve

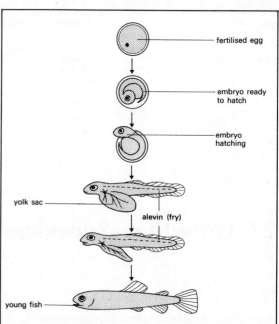

Figure 27.4 Development of trout

grows and moults several times. With each moult the nymph becomes more like the adult, which finally emerges to repeat the cycle.

Growth curves

Compared to that of the toad tadpole, the growth curve of a young grasshopper (figure 27.3) takes the form of a 'staircase'. Whereas the tadpole increases in size continuously, the insect's growth is periodically restricted by its rigid exoskeleton. Each such 'plateau' in the graph is however followed by a steep incline as the young insect undergoes moulting (**ecdysis**) and then rapid growth for a short time until once again its progress is hampered by its skin.

Trout

A fertilised egg (figure 27.4) contains a zygote and a large yolky food store. After a period of growth the embryo hatches from the egg. At first the young fish (**alevin**) feeds on the remainder of the **yolk sac** derived from the egg, but as this runs out it starts to catch its own food (pond insects) and eventually grows into an adult.

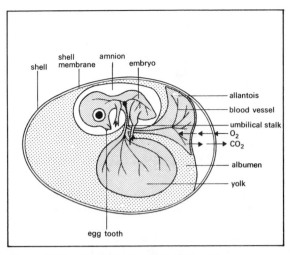

Figure 27.6   Development of chick embryo

Frog

Newly hatched tadpoles (figure 27.5) depend at first on the remaining egg yolk for food, then they eat plants and in the final stages of development consume small insects. Gaseous exchange initially occurs directly through the embryo's skin, then by **external gills**, then **internal gills** and finally by **lungs**. Since the young animal undergoes a series of changes before growing into the adult (e.g. absorption of tail into body) this is a further example of metamorphosis.

Bird

The ten day old chick embryo (figure 27.6) is surrounded, cushioned and therefore protected by the **amnion**, a fluid-filled sac. The **albumen** supplies it with protein and water; the **yolk** provides protein and fat. The **allantois** stores excretory products and transports oxygen to, and CO_2 away from the embryo in blood vessels. The permeable **shell** allows gaseous exchange with the atmosphere. Materials enter and leave the embryo via blood vessels which pass through the umbilical stalk.

After 21 days of **incubation** at about 40°C, the embryo fills the whole egg. Using its egg tooth it cracks the shell open from inside and hatches out.

Mammal

Following implantation of the fertilised egg(s) into the uterus wall, the period of development (**gestation**) begins. The embryo (e.g. human, figure 27.7) increases in size and complexity and when its species can be recognised from its appearance it is called a **foetus**. Like the chick it is protected by the amnion, but unlike the bird, the mammal's **umbilical cord** is attached to the **placenta**, an organ which

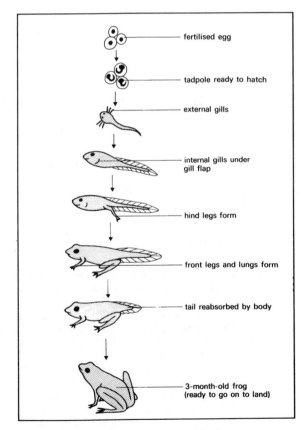

Figure 27.5   Development of young frog

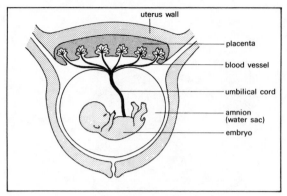

Figure 27.7   Development of human embryo

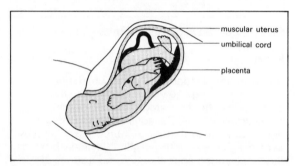

Figure 27.8   Birth

allows the maternal and foetal blood supplies to come into very close contact without them actually mixing. Here oxygen and CO_2 are exchanged, digested food, vitamins and minerals are added to the foetal bloodstream and excretory products are removed from it and returned to that of the mother.

After 40 weeks of gestation under these ideal conditions of protection and nourishment, the foetus is fully grown and ready to be born (figure 27.8). The muscular uterus begins to contract rhythmically ('**labour**' has begun). The amnion bursts, the baby is forced out head first and its umbilical cord is tied and cut. In response to the sudden drop in temperature, the infant takes its first breath and from then on uses its own lungs to obtain oxygen. Later the placenta is expelled as the **after-birth**.

Parental care

This is only found to any great extent in some vertebrates. Birds, for example, feed their chicks and protect them from predators and cold temperatures until they can fly. All mammals suckle their young with milk from the **mammary glands**. In addition they build a **nest** which along with the adults' body heat protects the young from cold temperatures and

stops them straying into the clutches of predators.

Parental care lasts much longer in human beings than in any other animal with the young receiving lengthy preparation and education for their future life as independent adults.

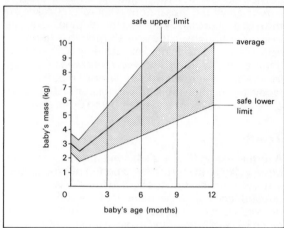

Figure 27.9   Human baby growth curves

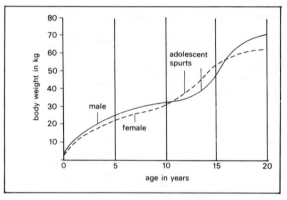

Figure 27.10   Adolescent spurt graph

Human development

Immediately after birth, a baby's weight drops briefly as it uses up some of its food reserves before gaining nourishment from the mother's milk. However soon its weight increases rapidly as shown in figure 27.9. The shaded area of the graph represents the limits within which a healthy baby's weight should remain. Weight above or below these limits indicates a health problem.

Although girls tend to be lighter and shorter than boys (figure 27.10), girls enter their **adolescent spurt** sooner than boys and therefore overtake them for a time. However boys soon catch up again and therefore adult men are taller and heavier on average than adult women.

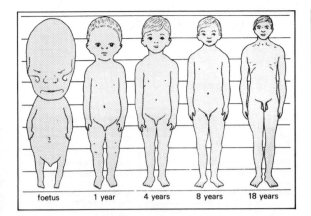

Figure 27.11 Differing growth rates of body parts

Growth and development of the human body from foetus to adult also involves changes in the body's proportions as shown by the various stages illustrated in figure 27.11.

Revision questions

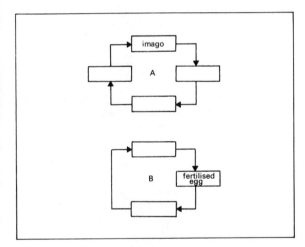

Figure 27.12 see question 1

1 **(a)** Figure 27.12 shows 2 types of insect life cycle. Redraw them and fill in the blank boxes using the following words:
   adult; fertilised egg; pupa; nymph; larva.
  (b) Identify the type of metamorphosis in each case.
2 Why does a young trout alevin whose mouth has not yet opened not die of starvation?
3 Beginning with the term fertilisation, rewrite the following in the correct order:
   adolescence; adulthood; birth; childhood; fertilisation; gestation; implantation.

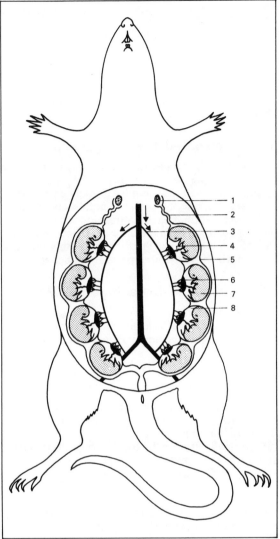

Figure 27.13 see question 4

4 **(a)** Name parts 1–8 in figure 27.13.
  (b) State the function of parts 1, 4, 6 and 8.
5 By referring to figure 27.9, find out which baby in the following table is **(a)** increasing in weight too rapidly **(b)** not gaining weight rapidly enough.

baby	weight in kg at birth	weight in kg at 6 months
A	3.6	8.7
B	2.4	3.7
C	3.7	8.4
D	2.5	3.4

28 Sexual reproduction in flowering plants

Flowers are the reproductive organs of an **angio-sperm** plant.

Insect-pollinated flower (figure 28.1)

Each **stamen** is a male organ which produces **pollen grains** (male gametes). A **carpel** is a female organ containing one or more **ovules** (female gametes) in an **ovary**. The brightly coloured **petals** which often have **nectar** at their bases, attract insects to the flower. **Sepals** protect the unopened floral bud. The **receptacle**, the swollen tip of the floral stalk, acts as a base to which all the other parts are attached.

Wind-pollinated flower (figure 28.2)

The male and female organs develop inside two leaf-like bracts. The basic differences between wind and insect-pollinated flowers are summarised in table 28.1.

Pollination

Self-pollination is the transfer of pollen from a stamen to a stigma in the same flower or in another flower on the same plant.

   Cross-pollination is the transfer of pollen from a stamen of one flower to a stigma in a flower on another plant of the same species.

Wind pollination

This is a wasteful method since many of the tiny pollen grains never reach a stigma.

Insect pollination

In this more efficient method, the pollen grains stick to an insect's body and are carried from flower to flower. Some flowers are elaborately adapted to increase the chance of insect pollination. When the flower in figure 28.3 is visited by an insect, its stigma is first to come into contact with the bee and receive pollen from the previous flower. As the bee probes for nectar, pollen grains from this flower stick to its body and are carried on to the next flower.

Fertilisation

A pollen grain absorbs the sugar solution on the sticky stigma and begins to germinate by forming a **pollen tube** (figure 28.4). The pollen grain's two nuclei pass into the tube and the first one disintegrates. The second divides into **two male**

feature	wind-pollinated flower	insect-pollinated flower
size of flower	small	large
petals	green bracts, no scent or nectar	colourful, scented, nectar often present
stamens	dangle out of flower (easily shaken by wind)	enclosed by petals (brushed against by insects)
pollen	many small pollen grains produced	fewer larger sticky pollen grains produced
stigma	feathery, outside flower (to catch pollen)	sticky, inside flower (to meet passing insect)

Table 28.1   Comparison of wind and insect-pollinated flowers

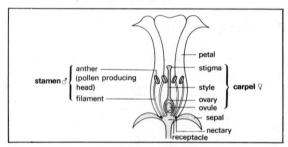

Figure 28.1   Insect-pollinated flower

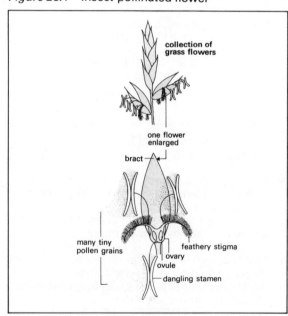

Figure 28.2   Wind-pollinated flower

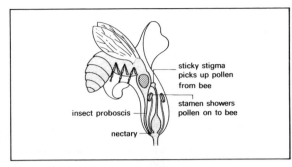

Figure 28.3 Insect pollination

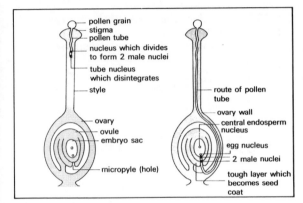

Figure 28.4 Fertilisation

gametes which occupy the tip of the tube as it grows down the style, round the ovary and into the ovule by a tiny hole (**micropyle**). **Internal fertilisation** now takes place as the first male gamete enters the **embryo sac** (figure 28.4) and fuses with the **egg** (ovum) **nucleus** forming the **zygote** (which later grows into the embryo plant). Fusion between the second male gamete and the **central endosperm nucleus** results in the formation of **endosperm** (food later used by the developing embryo).

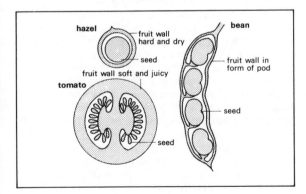

Figure 28.5 Fruits

After fertilisation the ovule (now an embryo plant and foodstore surrounded by a tough coat) is called the **seed**. The ovary (whose wall has now become hard and woody or soft and juicy, or dry and membranous etc.) is called the **fruit** (figure 28.5). It may contain one or more seeds.

Seed dispersal

Seeds and fruits are dispersed by several different agents (figure 28.6).

Good seed dispersal distributes a species over a wide range of habitat thus increasing its chance of survival and preventing intense competition for water, light, space etc. being set up between parent and young plants.

Wind
Some seeds bear **wings** (e.g. sycamore) or **para-chutes** (e.g. dandelion) which keep their airborne; others (e.g. poppy) are tiny and are scattered when the wind makes their '**pepper-pot**' container sway from side to side.

Animals
Some seeds (e.g. cherry) are surrounded by **juicy** tissue which is eaten by animals. The hard seed is spat out or passed undamaged through the animal's gut. Seeds (e.g. oak acorn) containing **fat** attract animals which hibernate. Some of the seeds are eaten and others are dispersed as the animals store them for future consumption. Other seeds bear **hooks** (e.g. goosegrass) which fasten on to the fur of passing animals.

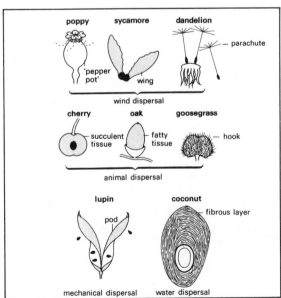

Figure 28.6 Seed dispersal

Mechanical (self dispersal)

When a pod (e.g. lupin) is ripe, it springs open and ejects the seeds.

Water

Some fruits are designed to float. A coconut, for example, is surrounded by a **fibrous layer** which contains many air spaces.

Revision questions

1 Unlike a sperm, a pollen grain has no tail with which to swim. How does it travel **(a)** from anther to stigma **(b)** from stigma to embryo sac?
2 **(a)** Name parts 1–8 in figure 28.7.
   (b) If no further pollination occurs, what is the maximum number of seeds that this fruit can contain?

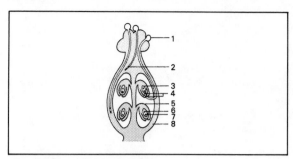

Figure 28.7   see question 2

3 State 4 differences normally found between wind and insect-pollinated flowers.
4 Which agent of pollination also acts as an agent of seed dispersal?
5 Stage 2 advantages of good seed dispersal.

29 Growth and development of flowering plants

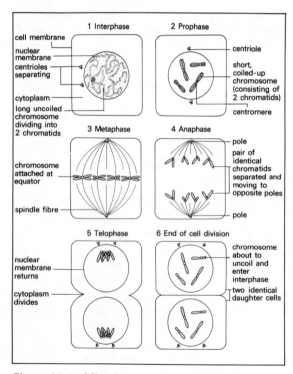

Figure 29.1   Mitosis

Formation of embryo plant

The unicellular **zygote** (formed at fertilisation) becomes a multicellular **embryo** plant by **cell division**. During this process, the nucleus divides first, followed by the cytoplasm so that each daughter cell receives half of the original cell's protoplasm.

Mitosis

Nuclear division (**mitosis**) involves several stages (figure 29.1). Following **interphase**, thread-like **chromosomes** appear. As each chromosome becomes shorter and thicker is is seen to be a double thread where each of the component threads is a **chromatid**. The two chromatids of each chromosome are joined together at the **centromere**. By **metaphase**, the nuclear membrane has disappeared, a **spindle** has formed and each chromosome is attached by its centromere to one of the spindle fibres at the **equator** (equatorial plate). During **anaphase** each centromere splits and one chromatid from each pair moves to the 'north' pole and one to the 'south' pole. During **telophase**, a nuclear membrane forms round each group of chromatids (now regarded as chromosomes) and the cytoplasm divides to produce **two identical daughter cells**.

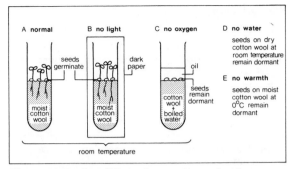

Figure 29.2 Conditions for seed germination

Germination

This is the development of the embryo plant in a seed into an independent plant with its first green foilage leaves. To germinate, seeds require **oxygen**, **water** and **warmth** as demonstrated by the experiment shown in figure 29.2. Only the seeds in tubes A and B germinate, so seeds grow in both light and dark.

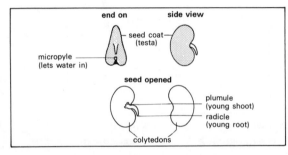

Figure 29.3 Structure of broad bean seed

Germination of broad bean

The structure of a broad bean seed is shown in figure 29.3. Broad bean is an example of a **dicotyledonous** plant because its seed has two cotyledons (storage leaves) which contain starch.

Radicle

Figure 29.4 shows the outcome of marking the radicle of a germinating broad bean seed with ink at 1 mm intervals and then allowing it to grow for a few days. The space between the tip and mark 1 remains unaltered since this region consists of the protective root cap and the region of mitosis. Marks 1, 2, 3 and 4 do, however, become very spaced out since the cells in this region of the root are undergoing **elongation** and **vacuolation**. Above mark 4 there is no change in the spacing of the ink marks because here the fully elongated cells are undergoing **differentiation**. This means that each is becoming structurally adapted to suit its future specialised role. For example some cells become long hollow tubes supported by lignin (xylem vessels); others become long thin projections out into the soil (root hairs).

Plumule

The plumule, at first curved to protect its delicate growing point as it pushes up through the soil, emerges as the young shoot. Until the first green foliage leaves appear (figure 29.5), the embryo plant is entirely dependent upon the cotyledons for food. Digestion of the stored starch to simple sugar is promoted by the enzyme **diastase** (plant amylase). The sugar is used up in tissue respiration to release energy for growth and the seedling shows a temporary decrease in dry weight.

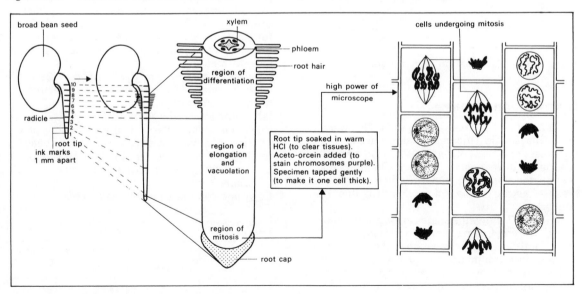

Figure 29.4 Growth of broad bean radicle

Once photosynthesis begins, the seedling grows into a mature plant which eventually bears flowers. These become pollinated, have their ovules fertilised and produce seeds which germinate and begin the cycle over again.

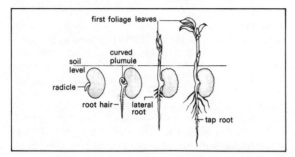

Figure 29.5   Germination of broad bean seed

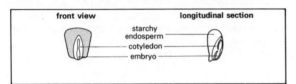

Figure 29.6   Structure of maize grain

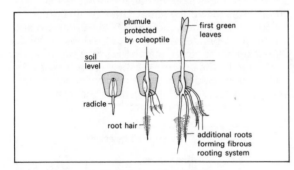

Figure 29.7   Germination of maize grain

Germination of maize

The structure of a maize grain is shown in figure 29.6. This plant is a **monocotyledon** since it possesses only one cotyledon. When iodine solution is added to the exposed surface of a longitudinal section, the starchy **endosperm** which turns blue-black is found to be in a region behind the shield-like cotyledon.

Figure 29.7 shows the stages in the germination of maize. The cotyledon absorbs food from the endosperm and passes it to the young root and shoot. The shoot grows straight up through the soil since its delicate young leaves are protected by a sheath, the **coleoptile**, which bursts open allowing the first green leaves to emerge.

Meristems

A **meristem** is a tissue whose cells are able to divide forming new tissues. At first the whole embryo is a meristem, but soon meristematic activity becomes restricted to the root and shoot tips and the cambium (see chapter 20).

plant	animal
growth only occurs at meristems	growth occurs all over body
growth (increase in size) occurs throughout life (indeterminate)	growth (increase in size) stops on reaching adulthood (determinate)

Table 29.1   Comparison of plant and animal growth

Revision questions

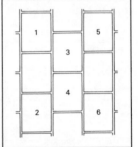

Figure 29.8
see question 2

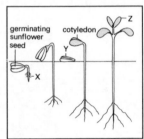

Figure 29.9
see question 3

1 Beginning with germination of the seed, rewrite the following stages in the life cycle of a plant in the correct order:
   fertilisation; flowering; germination of seed; growth of young plant; pollination; seed dispersal; seed formation.
2 Figure 29.8 corresponds to part of figure 29.4. Name the stage of mitosis occurring in each of the numbered cells.
3 **(a)** Name structures X, Y and Z in figure 29.9.
   (b) What additional role do the cotyledons play here compared with those of a broad bean?
   (c) In what way does germination of a sunflower seed differ from that of both broad bean and maize?
4 Copy and complete the following table:

condition required for seed germination	reason why it is necessary
	to provide optimum temperature for enzymes to act
	to oxidise food during aerobic respiration
	to act as the solvent in which metabolic reactions can occur

30 Asexual reproduction

When new individuals are produced by a single parent without involving gametes or fertilisation, such reproduction is said to be asexual and the genetically identical offspring are said to make up a **clone**.

Plants

Binary fission
This method of asexual reproduction is common amongst unicellular organisms such as bacteria and *Pleurococcus* (figure 30.1), in which a single cell divides to produce two **similar daughter** (uniform progeny) cells. Repeated binary fission results in the formation of large numbers of **uniform progeny**.

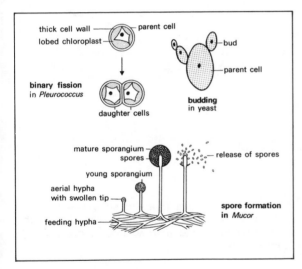

Figure 30.1 Asexual reproduction in simple organisms

Budding
In the unicellular fungus yeast, a new individual is formed as an outgrowth, a **bud** (figure 30.1), of the parent cell. As the bud increases in size, it may break off to lead an independent life or stay attached and form another bud leading eventually to a chain of cells.

Spore formation
The saprophytic fungus *Mucor*, consists of a tangled mass of branching threads called hyphae. Figure 30.1 shows how the tip of a reproductive hypha develops into a black **sporangium** containing hundreds of tiny **spores**. On release the spores are carried for great distances by the wind. A spore which happens to land on favourable conditions (e.g. moist bread in a warm room) will germinate into a new *Mucor* colony.

Natural vegetative propagation
In angiosperms this form of asexual reproduction involves the production of multicellular offspring by a parent plant.

Bulb
In spring, food stored in the fleshy leaf bases of a **bulb** (e.g. tulip, figure 30.3) passes into the bud which grows into this year's shoot. After flowering, food from this year's green leaves is stored in their fleshy bases which surround next year's bud. Thus the new bulb is formed **inside** the old one. Side buds similarly form daughter bulbs.

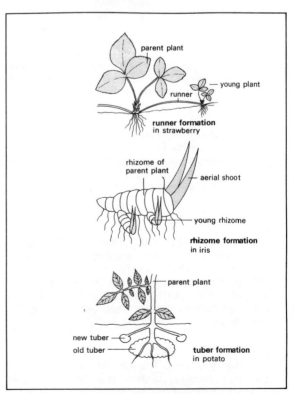

Figure 30.2 Asexual reproduction in angiosperms

Corm

A **corm** (e.g. crocus, figure 30.3) is a short stem full of stored food which passes up into the bud in spring. After this has developed into a shoot and flowered, food passes from the green leaves into the new corm which is formed **on top** of the old one. Side buds similarly form daughter corms.

daughter plants at some distance from the parent plant.

Artificial vegetative propagation

In angiosperms this method of asexual reproduction involves the detachment of one or more multi-cellular offspring from a parent plant by man.

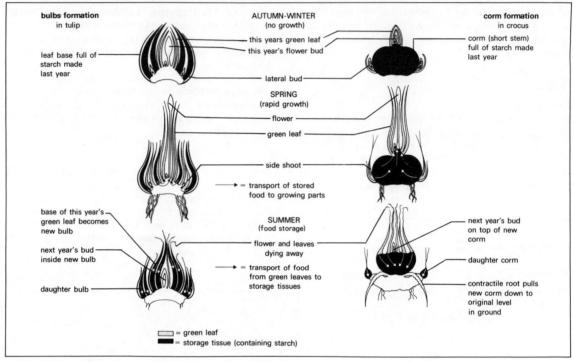

Figure 30.3   Asexual reproduction in angiosperms cont.

Rhizome

This is an **underground stem** which sends up an aerial shoot (e.g. iris, figure 30.2). In summer, food passes from the green leaves down to the main bud, which grows and continues the rhizome. Lateral buds become new rhizomes which branch out from the original.

Tuber

In a potato plant (figure 30.2), lateral buds produce underground shoots. The ends of these shoots swell up to form **tubers** full of food received in summer from the green leaves above the ground. The next year, one (or more) of the potato tuber's eyes (lateral buds) uses the food store and grows into a new potato plant.

Runner

In strawberry, some of the side (lateral) buds on the parent plant produce **runners** (figure 30.2). These are shoots which grow horizontally over the ground. Their terminal buds produce roots and become

Cutting

When the cut end of a geranium shoot is place in water, it develops roots which grow directly from the stem and continue to grow when the cutting is planted in soil.

Grafting

A cutting (**scion**) from a cultivated (and often

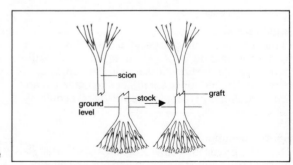

Figure 30.4   Grafting

delicate) variety of rose bush or fruit tree is grafted on to the **stock** of a hardy (and often disease-resistant) variety (figure 30.4). Cell division in the cambium at the two cut surfaces produces tissue which bonds them together healing the wound. Grafting is necessary because, on its own, the cultivated variety would not grow well. However the stock often develops wild shoots from points below the graft. These 'suckers' must be cut off (pruned) to prevent them from depriving the 'mixed' plant's upper part of water and essential minerals.

Animals

Asexual reproduction only occurs in very simple, primitive animals such as the protozoan, *Amoeba*, which divides by binary fission and the coelenterate, *Hydra*, which can produce new offspring by budding.

Revision questions

1 (a) Match each organism in the following group with the type of reproductive unit that it makes:
crocus, spore, daffodil, rhizome, iris, corm, *Mucor*, bulb.
(b) What reproductive unit is also made by a crocus plant by sexual reproduction?
2 By which of the following two methods is **(a)** a new variety formed **(b)** more of an existing variety formed?

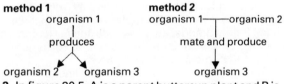

method 1
organism 1
produces
organism 2   organism 3

method 2
organism 1——organism 2
mate and produce
organism 3

3 In figure 30.5, A is a parent buttercup plant and B is a young plant produced by asexual reproduction.
(a) Name structure C.
(b) What name is given to this group of genetically identical buttercup plants?
(c) State one disadvantage of vegetative reproduction.

asexual	sexual
advantages offspring genetically identical to parent (useful when increase in a certain plant type required) young plant often possesses a food store (allowing rapid early growth and flowering)	**advantages** some characteristics inherited from one parent some from the other, thus great variation amongst offspring (whatever occurs in environment e.g. disease there is a good chance that some will survive) offspring well distributed
disadvantages no variation (environmental change e.g. disease may wipe out entire population) offspring not well distributed (therefore compete for space, light, water etc.)	**disadvantage** depends on the meeting and fusion of two gametes which cannot always be ensured (animals can search for a mate but flowers depend on wind or insects for pollination)

Table 30.1   Comparison of sexual and asexual reproduction

4 Figure 30.6 shows a further method of grafting.
(a) Using the terms stock and scion, identify parts X and Y.
(b) Explain why this process is best done in spring.

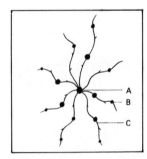

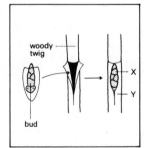

Figure 30.5
see question 3

Figure 30.6
see question 4

31 Variation

Although the members of a species of animal or plant are normally very similar to one another (e.g. a population of cats all possess typical cat-like features), they are not identical to one another since **variation** exists among them (e.g. in a cat population some animals will be bigger, some fatter, some lighter in coat colour etc.) Similarly although all of the leaves shown in figure 31.1 belong to the one species of plant (*Taraxacum officinale*, the dandelion), they vary in size and shape.

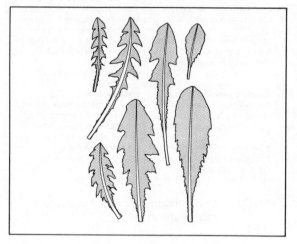

Figure 31.1 Variation amongst dandelion leaves

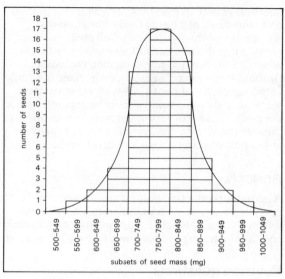

Figure 31.2 Continuous variation in mass of seeds

seed number	mass (mg)	seed number	mass (mg)	seed number	mass (mg)
1	597	21	751	41	811
2	622	22	755	42	813
3	648	23	758	43	819
4	661	24	762	44	823
5	672	25	765	45	828
6	683	26	768	46	830
7	699	27	769	47	835
8	711	28	770	48	837
9	718	29	770	49	841
10	723	30	774	50	843
11	725	31	777	51	846
12	727	32	782	52	848
13	735	33	785	53	850
14	739	34	787	54	862
15	741	35	791	55	871
16	741	36	793	56	885
17	743	37	799	57	897
18	745	38	800	58	923
19	746	39	801	59	947
20	748	40	806	60	952

Table 31.1 Variation in mass of seeds

Types of variation within a species

Continuous variation
A characteristic is said to show **continuous variation** if it varies in a smooth continuous way from one extreme to the other. Examples in animals include body height, handspan, pulse rate and tail length and in plants include seed mass, number of fruits on plant and number of leaves on a tree.

Since there are no clear cut dividing lines between varieties, the entire range of the characteristic is divided, for convenience, into many small categories (**subsets**). This allows a **histogram** (figure 31.2) to be plotted. Most individuals fall near the centre of the range with fewer at the extremities and when a curve is drawn, a bell-shaped **normal distribution** is found to be the result. In figure 31.2, the **range** in mass of the sixty castor oil seeds listed in table 31.1 is found to extend from 550 mg to 999 mg and the **most common** seed mass is the subset 750 to 799 mg.

Discontinuous variation
A characteristic is said to show **discontinous variation** if it falls into two or more **distinct** categories (the organism either has the characteristic or has not). Examples include blood group, presence or absence or ear lobes (figure 31.3), eye colour, fingerprints (figure 31.4), green or variegated leaf etc.

When presented as a graph, each group is kept separate from the others. Figure 31.5 shows an analysis of 100 forefinger prints.

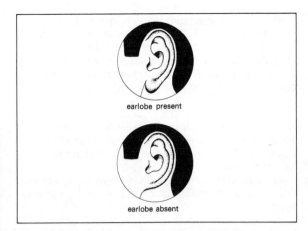

Figure 31.3   Discontinuous variation in ear types

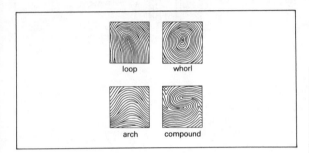

Figure 31.4   Discontinuous variation in fingerprint types

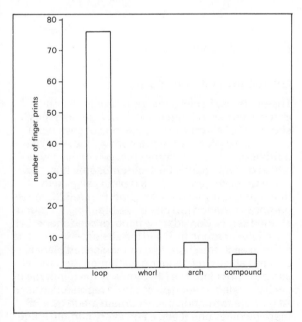

Figure 31.5   Histogram of discontinuous variation

Genetic control

Characteristics which are passed on from generation to generation are controlled by factors called **genes** (see chapter 32). Characteristics which show continuous variation are influenced by both inherited factors and the environment. For example a person may inherit the genes governing tallness but receive an inadequate diet during childhood and therefore not reach his full potential height. Characteristics which show discontinous variation are controlled solely by genetic factors. A person's blood group, for example, is established genetically and remains constant throughout life irrespective of environmental factors.

Revision questions

1 (a) Explain the meaning of continuous variation.
  (b) Which of the following is an example of discontinuous variation?
       Waist circumference, male or female sex, state of teeth, breadth of foot.

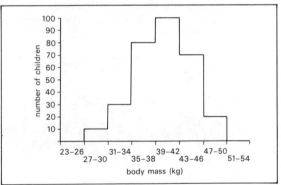

Figure 31.6   see question 2

2 Figure 31.6 shows variation in body mass of a population of schoolchildren.
  (a) Does body mass vary continuously or discontinuously?
  (b) How many children were weighed?
  (c) What was the most common body mass?
  (d) What was the range in body mass?
  (e) Redraw the graph and then draw in the appropriate normal distribution curve.
  (f) On the same diagram draw a second curve to represent a similarly-sized population of undernourished children living in famine conditions.

3 Under the headings continuous and discontinuous variation, list 4 ways in which identical twins reared in different environments (a) would definitely be exactly alike and (b) could differ from one another.

32 Genetics

Mendel's experiments

Mendel, an Austrian monk, was first to put genetics (the study of heredity) on a firm scientific basis. He carried out breeding experiments using several varieties of pea plants which possessed characteristics showing discontinuous variation (see chapter 31).

A plant is true (pure) breeding if, when self-pollinated, it produces more plants of the same type. In the experiment shown in figure 32.1, Mendel crossed pea plants which were true-breeding for production of round seeds with pea plants which were true-breeding for production of wrinkled seeds. All the seeds produced (the **first filial generation, F₁**) were found to be round. When these seeds had grown into plants, they were self-pollinated. The resultant **second filial generation (F₂)** contained some round seeds and some wrinkled in the ratio of 3 round : 1 wrinkled.

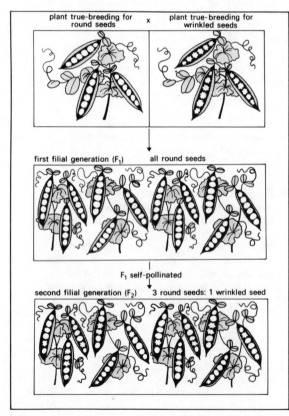

Figure 32.1 One of Mendel's experiments

Wrinkled seed absent in the F₁ generation had reappeared in the F₂, so 'something' had been transmitted undetected in the gametes from generation to generation. Mendel called this a factor. To-day we call it a **gene**. The gene in the above example controls the characteristic of seed shape in pea plants. There are various forms (**alleles**) of this gene. One expression (allele) produces round seeds, another wrinkled. Since the presence of the round allele masks the presence of the wrinkled allele, the round allele is said to be **dominant** and the wrinkled one **recessive**.

The appearance of an individual with respect to a particular inherited characteristic is called its **phenotype**.

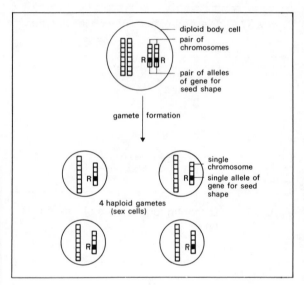

Figure 32.2 Gamete formation

Genes and chromosomes

The nucleus of a cell contains a definite number of **chromosomes**. A **haploid** cell (e.g. a gamete) has a **single** set of chromosomes (i.e. one of each type) whereas a **diploid** cell (e.g. a normal body cell) has a **double** set of chromosomes (i.e. two of each type which occur in pairs). Each chromosome bears one allele for each gene and in a diploid cell exactly matches its partner gene for gene. Figure 32.2 shows gamete formation in a plant true-breeding for round-shaped seeds. Consider the gene for seed shape. Let the allele for round be **R** and the allele for wrinkled be **r**. Since each true-breeding round-seeded plant has 2 round alleles, its genetic constitution (**genotype**) can be represented as **RR**. Similarly the genotype of each wrinkled-seeded plant can be represented as **rr**. At gamete formation every gamete made by a **RR** plant receives one **R** allele and every gamete made by a **rr** plant one **r** allele.

The cross can be represented as follows.

original cross	RR	× rr
gametes	all R	all r
F₁ genotype	all	Rr
phenotype	all	round
second cross	Rr	× Rr (self-pollinated)
gametes	R and r	R and r

pollen

	R	r
R	RR	Rr
r	rR	rr

F₂ genotypes ovules

F₂ phenotypic ratio 3 round : 1 wrinkled

When an individual possesses 2 similar alleles of a gene (e.g. R and R or r and r) its genotpye is said to be **homozygous** and all of its gametes are similar with respect to that characteristic.

When an individual possesses 2 different alleles of a gene (e.g R and r) its genotype is said to be **heterozygous**. It produces 2 different types of gametes with respect to that characteristic.

Mendel was successful because he studied one characteristic at a time and worked with large numbers of plants. In the above experiment, the F_2 generation consisted of 5474 round and 1850 wrinkled seeds. This is not an exact 3 : 1 ratio owing to the random nature of pollination and fertilisation which both involve a certain degree of chance. In the light of several similar experiments all producing an F_2 ratio of approximately 3 : 1, Mendel developed a theory to explain the facts. Expressed in modern terms it states that:

The inheritance of characteristics is determined by genes.
The alleles of a gene exist in pairs.
At gamete formation each gamete only receives one of a pair of alleles.
Alleles retain their identity from generation to generation.

Use of fruit flies in genetic experiments

Two features of the reproduction of fruit fly (*Drosophila melanogaster*) that makes it suitable for the study of genetics are the shortness of its life cycle (about 10 days) and the vast number of offspring produced (which allows valid statistical analysis of the results). Figure 32.3 shows the difference between male and female fruit flies.

Inheritance of wing type in *Drosophila*
In the cross about to be considered, normal-winged males are crossed with vestigial-winged females

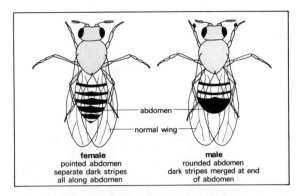

Figure 32.3 'Wild-type' fruit flies (i.e. showing normal features)

Figure 32.4 Vestigial-winged fruit fly

(figure 32.4). However in terms of genetics, a cross between normal-winged females and vestigial-winged males would give exactly the same results.

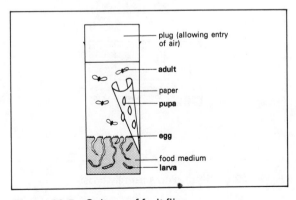

Figure 32.5 Culture of fruit flies

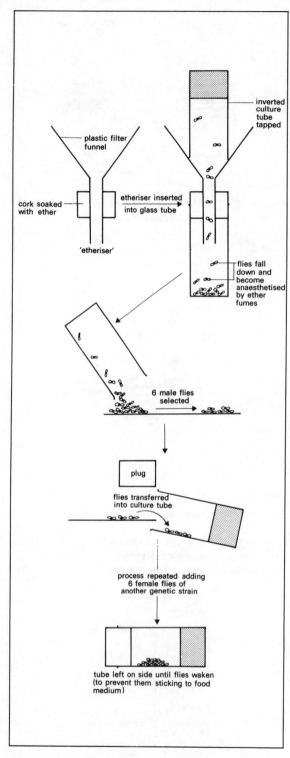

In any cross the females must be virgins to ensure that the eggs have not been fertilised in advance by a fly of unknown genotype since this would clearly invalidate the results.

Figure 32.5 shows a culture tube which provides everything a fruit fly needs in its life cycle. Figure 32.6 shows the method used to set up a cross. Several of each sex are used to allow for non-recovery from the anaesthetic. After egg laying, the parents are discarded and the F_1 generation allowed to develop at 25°C. When the adult flies appear, a few are transferred to new culture tubes by the same method and left to produce the F_2 generation. The results obtained are summarised as follows.

original cross	normal wing	× vestigial wing
F_1	all normal wing	
second cross	normal wing	× normal wing (F₁ self-fertilised)
F_2	3 normal wing : 1 vestigial wing	

In *Drosophila*, dominant traits are often represented by the symbol + and recessive traits by their own first letter(s). Consider the gene for wing type. Let the normal wing allele (which the above results show to be dominant) be + and the recessive vestigial wing allele be vg. Then the cross can be represented as follows.

original cross	++	× vgvg
gametes	all +	all vg
F_1 genotype	all +vg	
phenotype	all normal wing	
second cross	+vg	× +vg (F₁ self-fertilised)
gametes	+ and vg	+ and vg

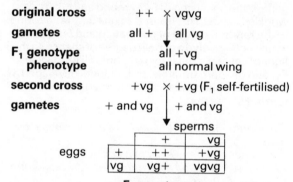

		sperms	
		+	vg
eggs	+	++	+vg
	vg	vg+	vgvg

F_2 genotypes

F_2 phenotypic ratio 3 normal wing : 1 vestigial wing

Backcross

When an organism exhibits a dominant trait, it is not obvious whether its genotype is homozygous or heterozygous for that trait. The identity of an unknown genotype can be found by backcrossing it with a homozygous recessive organism as shown in figure 32.7.

Human genetics

The genetic relationships with respect to a certain gene are often presented as a family tree. Figure 32.8

Figure 32.6 Setting up a fruit fly cross

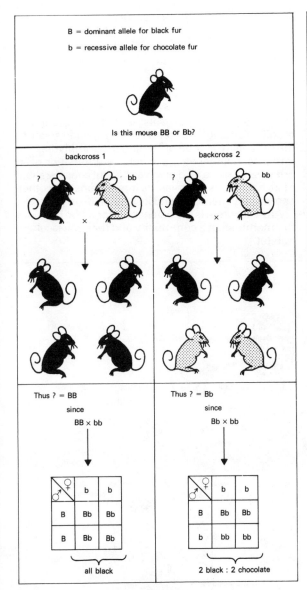

B = dominant allele for black fur

b = recessive allele for chocolate fur

Is this mouse BB or Bb?

backcross 1	backcross 2

Thus ? = BB
since
BB × bb

all black

Thus ? = Bb
since
Bb × bb

2 black : 2 chocolate

Figure 32.7 Backcrossing

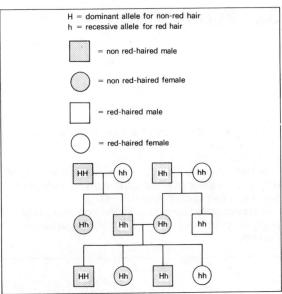

H = dominant allele for non-red hair

h = recessive allele for red hair

= non red-haired male

= non red-haired female

= red-haired male

= red-haired female

Figure 32.8 Family tree

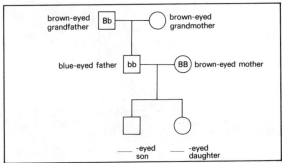

Figure 32.9 see question 3

shows the inheritance of the recessive trait red hair in a human family.

Revision questions

1 State the genetic term defined by each of the following statements.
   (a) the appearance of an organism
   (b) the genetic constitution of an organism
   (c) carrying 2 identical alleles for a gene
   (d) carrying 2 contrasting alleles for a gene

2 In fruit fly the allele for grey body (+) is dominant to ebony body (e). Copy and complete the following table.

parents	genotypes: +e × +e	
	phenotypes: grey | grey	
	gametes: and | and	
next generation	genotypes:	
	phenotypes:	
	phenotypic ratio	

3 In humans brown eye colour (B) is dominant to blue eye colour (b).
   (a) Copy and complete the family tree shown in figure 32.9.
   (b) Name 2 individuals who have the same phenotype but different genotypes with respect to this gene for eye colour.
   (c) If the son marries a blue-eyed woman, what chance is there of each of their children being brown-eyed?

99

33 Ecosystems

Ecology is the study of living organisms in relation to their environment.

A **habitat** is the place where an organism lives (e.g. lake, ocean, desert, loam soil, mountain top, bark of tree etc.)

A **population** is a group of organisms normally of the same species (e.g. a pack of wolves, a shoal of herring, a forest of conifer trees etc.)

A **community** consists of all the populations of plants and animals living together in a particular habitat.

The (ecological) **niche** of a species is its status, in relation to the other organisms, within the community. Although this term is often applied solely to the organism's mode of feeding (e.g. omnivore, tertiary consumer etc.), accurate determination of an organism's niche also involves careful consideration of its habitat, its enemies, its interactions with other community members and its way of life in general.

An **ecosystem** is the balanced interaction between the members of a community and their physical habitat.

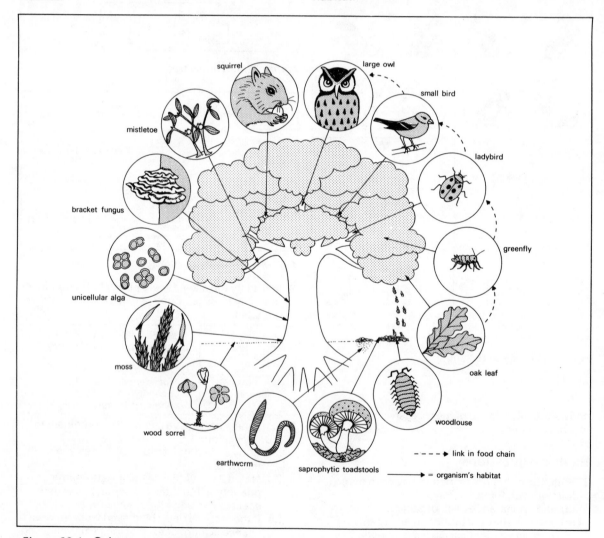

Figure 33.1 Oak tree ecosystem

Ecosystems

The nature of an ecosystem is determined by both non living (**abiotic**) and living (**biotic**) factors.

Abiotic factors

Temperature, rainfall, sunshine, type of soil etc. all play a part in determining which plants, and subsequently which animals, can survive in the habitat.

Biotic factors

An ecosystem's community consists of producers, consumers and decomposers. Energy passes along food chains and many feeding relationships such as mutualism, parasitism and saprophytism exist. Many other forms of **interaction** also occur such as **competition** between plants for light, space and water and **dependence** of animals on plants for food and shelter and of plants on animals for pollination and seed dispersal.

Interactions in an oak tree ecosystem

In an ecosystem, the most prevalent and influential member of the community is normally referred to as the **dominant** species. In the ecosystem shown in figure 33.1, this role is played by the oak tree. To thrive, this plant must receive sunlight, sufficient moisture and minerals and moderate temperatures. The soil type is also important. One species of oak, for example, prefers alkaline soil, another slightly acidic soil. Once established, the oak tree directly or indirectly affects the lives of all the other members of the community.

Plant community

In summer the dense shade cast by the thick canopy of oak leaves overhead prevents colonisation of the forest floor habitat by all but the specially adapted **shade-tolerant** plants (e.g. wood sorrel). These grow most in spring when they also flower ensuring seed production before their growth rate is curtailed by the shade in summer.

The tree trunk may be the habitat of other **producers** (e.g. mosses and unicellular algae) or it may play host to the **parasite**, bracket fungus, which obtains all of its nourishment from the tree.

The tree's branches may bear the **partial parasite**, mistletoe. This plant draws water and mineral salts from its host, but makes its own organic food by photosynthesis. It depends on birds to disperse its sticky seeds.

Animal community

Herbivorous insects (e.g. greenfly and caterpillars) which depend on the tree for food and shelter are eaten by **carnivorous insects** (e.g. ladybirds) and **small birds**. These consumers are in turn preyed upon by **predators** (e.g. owls) which also depend on the tree for a place to nest.

Squirrels shelter and hide from enemies in the tree and eat the acorns. By losing some during collection, they act as agents of seed dispersal and therefore return some benefit to the tree.

Dead leaves, discarded by deciduous trees in autumn, provide food for **earthworms** which in turn aerate the soil (providing oxygen for respiring roots), enrich the soil and turn it (see figure 34.14).

Decomposers

Dead leaves and wood also provide food for the forest's **decomposers** such as saprophytic toadstools and micro-organisms (bacteria and moulds) which bring about **partial** decay of dead remains forming **humus**.

Finally the producers depend on further **soil micro-organisms** for the **complete** breakdown of this dead organic matter and ultimately the release into the soil of **mineral salts** essential for healthy plant growth.

Interactions in a spruce tree ecosystem

Coniferous trees such as spruce thrive in cold wet climates. Unlike deciduous trees, they cast their leaves (**needles**) all the year round. The needles form a tough dark brown layer of raw **acidic humus** in which few saprophytic micro-organisms survive. Decomposition and subsequent release of minerals is therefore very slow. Even those plants that can tolerate low pH and shortage of nutrients are excluded almost without exception from this inhospitable habitat by the **evergreen leaf canopy**. It shuts out sunlight all the year round. Thus compared to oak trees with their quickly humified leaf litter and large interdependent community, spruce trees are poor supporters of plant and animal life.

Interactions in a moorland ecosystem

A moorland only forms under extreme climatic conditions of low temperature, high rainfall and exposure (especially in basin-shaped areas high up in mountainous regions where underlying material prevents drainage of water). These conditions lead to **poorly aerated** soil unable to support aerobic humus-forming bacteria. Thus when **sphagnum**, the bog moss, which does thrive in such extreme conditions, dies, it does not become decomposed. Instead layer upon layer of dead moss accumulates for hundreds of years forming **peat**.

Moorland community (figure 33.2)

Heather thrives on the somewhat drier areas of the acidic peaty soil and its young shoots supplement the insect diet of **red grouse** which is in turn preyed

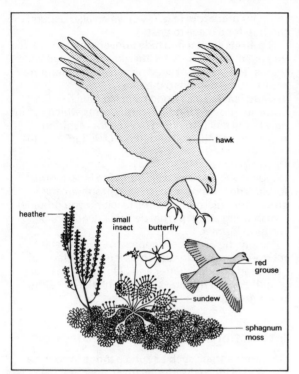

Figure 33.2 Moorland ecosystem

upon by **hawk**. Heather's nectar-rich flowers attract **insects** which receive food and in turn pollinate the flowers. The insects may however be caught and digested by the sticky red leaves of the insectivorous plant, **sundew**, which compensates for lack of **nitrogen** in the soil by taking the nitrogen from insects.

Mankind's effect on ecosystems

All the organisms in a stable ecosystem interact and interrelate with one another. However this harmonious and delicate balance of nature is often dramatically disturbed by man. Clearing of woodland for agriculture may lead to an increase in the number of wild grazing animals which in turn destroy the crops. Intense clearing of wooded hillsides may lead to erosion of fertile top soil by wind and rain. Hunting and killing foxes may disturb the predator-prey balance and cause an increase in mice and vole populations leading to severe crop damage. Building a dam to supply water for desert irrigation may mean permanently converting a rich valley into a lake.

Thus man's effects on an ecosystem are often enormous since he may not simply alter it but actually wipe it out entirely. The most serious of man's effects on this planet however is probably **pollution** (chapter 37).

Micro-habitats and micro-climates

Within large ecosystems communities of small animals are often found in **micro-habitats** such as under large stones or inside holes in trees. Here they are safe from predators and live in conditions of fairly constant **micro-climate** unavailable in the less sheltered places outside. Woodlice, for example, congregate under dead logs where the conditions are dark and damp.

Performing a biological investigation

In biology, the results of an investigation are valid if:
1. at each stage only **one variable factor** is studied at a time (if several are involved then it is impossible to know which was responsible for the results).
2. many organisms are used (if only a few are used then perhaps they were unusual, neurotic, tired, hungry, under stress etc.).
3. the experiment is successfully **repeated many times** (if not then perhaps the outcome just happened to result from a lucky chance).

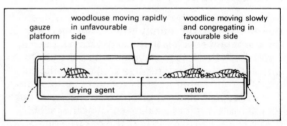

Figure 33.3 Choice chamber experiment

Investigating the effect of micro-climatic factors

A **choice chamber** is a piece of apparatus which allows one variable factor to be studied at a time. The one shown in figure 33.3 is used to investigate how woodlice respond to conditions of varying **humidity**. Obeying the three rules above, it is found that in this experiment the animals move in a **random** fashion, but move more **quickly** in dry **unfavourable** conditions and more **slowly** in damp **favourable** conditions where they tend, therefore, to congregate.

Similarly woodlice move quickly in **light** and slowly in **dark** where they tend to gather. Thus by testing one variable factor at a time it can be seen that woodlice respond to varying conditions of both humidity and light. This behaviour is of **survival value** to woodlice because in dry conditions they lose water through their permeable exoskeletons and die; and when in light, they are easily spotted and eaten by predators.

Population estimation

It would be very time-consuming to count every individual member of a population in an ecosystem (e.g. the dandelion plants growing in a large lawn). Instead several **quadrats** (figure 33.4) are thrown at **random** and the number of dandelions present in each quadrat counted. From this the average number of dandelions per square metre is calculated. The total area of the lawn in square metres is measured (length × breadth) and finally an estimate of the total population of dandelions calculated.

Revision questions

1 Copy and complete the following table using the answers: toadstool; sugary sap; leafy twigs; leaf litter; long-eared owl; greenfly.

member of community	food	habitat	ecosystem
greenfly			
	small bird	strong branches	oak tree
ladybird		trunk and branches	
		dead leaves	

2 (a) Draw a food web of the moorland community in figure 33.2.
   (b) Which community member is an omnivore?
3 Briefly explain why
   (a) all the woodlice used in the experiment in figure 33.3 must be of the same species,
   (b) there is an almost total lack of animals in a coniferous tree ecosystem.
4 The lawn in figure 33.4 is 11 metres in breadth and 14 metres in length. Estimate the total number of dandelions growing on it.

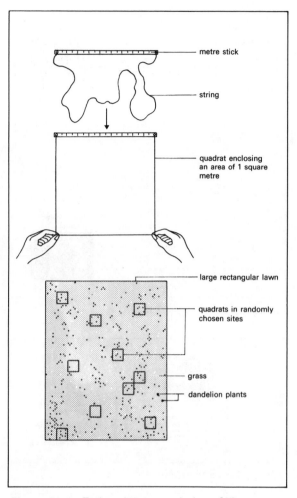

Figure 33.4 Estimating a population of plants

34 Soil

Soil formation

Weathering of **parent rock** occurs when rain water trapped in rock crevices expands on freezing and breaks up the rock. Hot weather causes the rock surface to expand and break away from the cooler underlying layers. **Lichens**, the first plants to colonise bare rocks, produce **acids** that gradually make crevices in the rocks allowing mosses and later small flowering plants to grow too. The dead bodies of these plants become converted by the action of micro-organisms into dark brown sticky **humus**. The **soil solution** gains its supply of **mineral salts** in the two ways shown in figure 34.1. After many years a mature soil with a **soil profile** (figure 34.2) is formed.

103

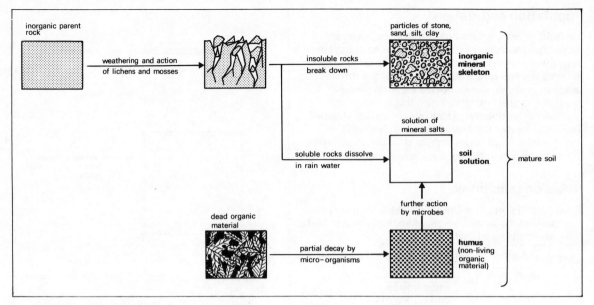

Figure 34.1 Soil formation

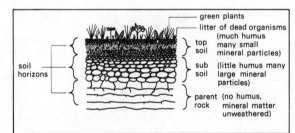

Figure 34.2 Soil profile

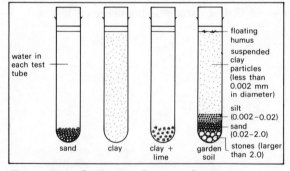

Figure 34.3 Sedimentation experiment

Composition and properties of soil

Solid components

Figure 34.3 shows the results of **sedimentation** tests, where a small sample of each soil is shaken up in water and allowed to settle. Unlike sand, clay particles are too small to gather as sediment at the bottom of the tube. However when **lime** (or barium sulphate) is added, the tiny clay particles stick together (**flocculate**) forming larger aggregates (particles) which do settle out. Loam (garden) soil contains a balanced mixture of different types of mineral particles.

Water content

100 g of fresh soil is dried in an oven at 95°C for 2 days and its new mass noted. This process is repeated until the sample reaches **constant mass** showing that all of its water has been removed.

100 − mass of dry soil = % **water content** of soil

Sandy soil has a low water content, loam a medium and clay a high water content.

Humus content

100 g of dry soil is roasted in a crucible for a few hours until it reaches constant mass.

100 − mass of roasted soil = % **humus content** of soil

Air content

This is measured by the method in figure 34.4. Clay soil contains less air than sand because its particles are smaller (figure 34.5).

Capillarity

Water rises furthest up the narrowest tube (figure 34.6). This process, by which water is drawn up through a narrow space, is called **capillarity**. Similarly in dry soil, the air spaces act as a series of irregular tubes (figure 34.7) allowing capillarity to occur. The smaller the air spaces (i.e. clay soil) the greater the capillarity.

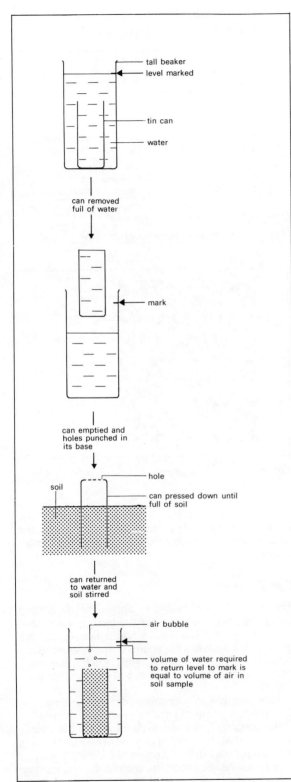

Figure 34.4   Measuring air content of soil

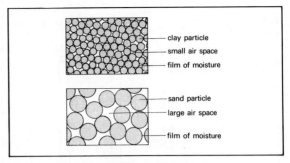

Figure 34.5   Sizes of air spaces

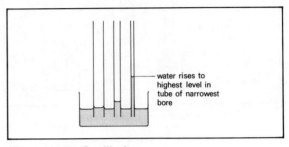

Figure 34.6   Capillarity

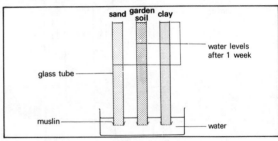

Figure 34.7   Soil capillarity

Permeability to air

This is the ease with which air passes through the soil. In figure 34.8 when the taps are opened, water runs out of the tube below sand fastest, showing that sand with its large air spaces is most **permeable to air**. The reverse is true of clay.

Permeability to water

When an equal volume of water is added to each soil sample shown in figure 34.9, most water quickly passes through the sand which, with its large particles and air spaces, is the most **permeable to water**. The reverse is true of clay which, with its tiny clinging particles and small air spaces, has the greatest power of **water retention**.

pH of soil

This is tested using the apparatus in figure 34.10. Fertile soils tend to be **neutral** since most crops do

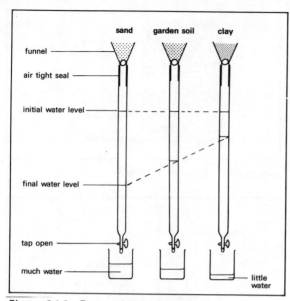

Figure 34.8　Permeability of soil to air

sandy soil	clay soil
large particles particles easily split up large air spaces – porous	small particles particles cling together small air spaces – not porous
poor water retention – good drainage mineral salts easily washed out (**leached**)	good water retention – poor drainage mineral salts not leached out
low capillarity warm and dry light to cultivate	high capillarity cold and wet heavy to cultivate
improved by adding clay or humus	improved by adding sand or lime

Table 34.1　Comparison of sandy and clay soil

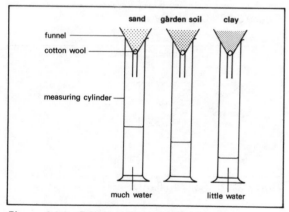

Figure 34.9　Permeability of soil to water

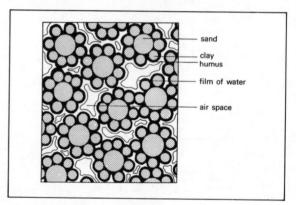

Figure 34.11　Soil crumbs

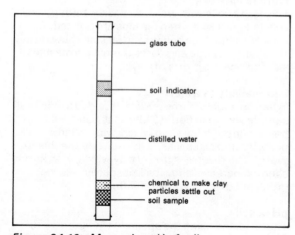

Figure 34.10　Measuring pH of soil

not thrive at high or low pH values. Soils with a low pH can be improved by adding **lime** to neutralise the **acid**. Moorland plants however are adapted to the acidic soil.

Fertile loam

The comparison in table 34.1 shows that sandy and clay soils are opposites each having good and bad characteristics. A **loam** soil tends to gain most of the good characteristics because it is a balanced mixture of sand and clay. These mineral particles become stuck together by humus into **soil crumbs** (figure 34.11).

This crumbly soil texture contains **air spaces** which allow adequate drainage which prevents water-logging (common in clay soil). On the other hand, the water-retaining powers of clay and humus in loam prevent drought (common in sandy soil). The air spaces between the crumbs also guarantee an **oxygen** supply for respiring roots and soil organisms.

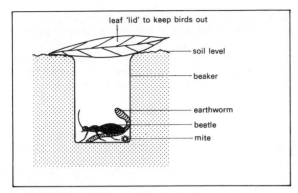

Figure 34.12 Simple trap

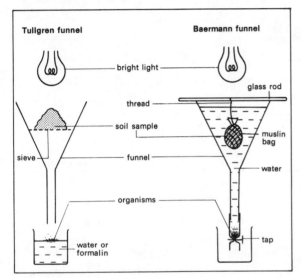

Figure 34.13 Funnel traps

Life in the soil

Traps

The simple trap shown in figure 34.12 is found to catch earthworms and animals such as beetles, centipedes and woodlice which live in leaf litter.

Funnels

A **Tullgren** funnel (figure 34.13) is used to extract tiny animals that live in the soil's air spaces e.g. thrips and springtail (figure 1.4 J). These animals move downwards away from the **bright, warm, dry** conditions created by the electric light bulb and fall through the sieve.

A **Baermann** funnel (figure 34.13) extracts tiny **aquatic** animals that live in the soil solution e.g. rotifer and nematode worm. Again these move downwards away from the bright, warm conditions and a small sample can be taken for inspection under the microscope by opening the tap.

Micro-organisms

Fertile soil contains a vast variety of **micro-organisms**. When a plate of nutrient agar is inoculated with a little fresh soil, profuse fungal and bacterial colonies grow (see chapter 36).

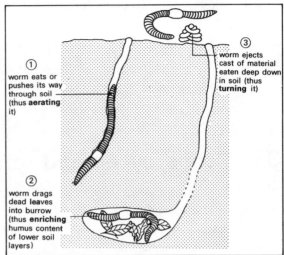

Figure 34.14 Effect of earthworms on soil ecosystem

Soil as an ecosystem

The physical components of soil act as a **habitat** by providing food and shelter for a vast **community** of living organisms. Since these living things both interact with one another and affect the state of their habitat (earthworms, for example, improve soil fertility, see figure 34.14), by definition, soil is an excellent example of an **ecosystem**. For its healthy maintenance this ecosystem is especially dependent on those bacteria which play certain crucial roles in the carbon, nitrogen and mineral salt cycles (see chapter 35).

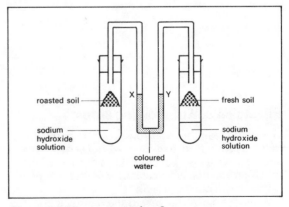

Figure 34.15 see question 3

Revision questions

1 Match the 3 'soil' samples in the following table with the locations:
country garden; sandy beach; peat bog.

'soil' sample	humus content	pH
1	70%	4
2	15%	7
3	1%	7

2 What chemical can be used to improve both moorland and clay soil? Briefly describe how it exerts its effect in each case.
3 Predict with reasons what will happen to levels X and Y in the experiment shown in figure 34.15.
4 State 6 differences between a clay and a sandy soil.

35   Crops, cycles and sprays

The maintenance of a balanced ecosystem depends on the repeated **cycling** in nature of the chemical elements **carbon** (figure 35.1) and **nitrogen** (figure 35.2). Similarly, **mineral elements** must be recycled as shown in figure 35.3.

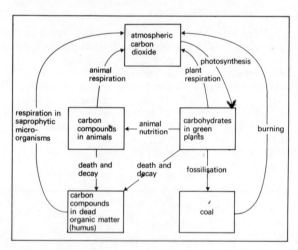

Figure 35.1   Carbon cycle

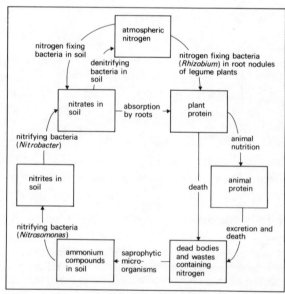

Figure 35.2   Nitrogen cycle

Effects on soil of growing crops

In order to feed the evergrowing world population, man has been forced to clear natural ecosystems and grow vast areas of crop plants such as wheat, potatoes and rice which absorb chemical elements from the soil during growth.

When the crop is harvested and removed from the ecosystem, the natural cycles are broken because no dead plant remains are returned to the soil. Mineral salts are no longer released from humus by micro-organisms and the soil becomes less and less fertile as the process is repeated again and again with the same crop.

Attempts to solve the problem

Fertilisers

Natural fertiliser (e.g. **manure**) is ploughed in during autumn to allow time for micro-organisms to release

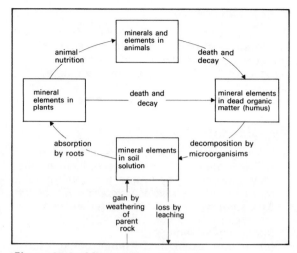

Figure 35.3 Mineral element cycle

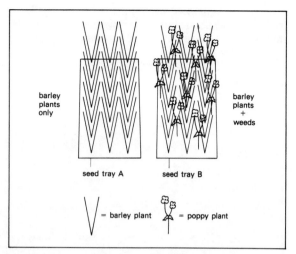

Figure 35.4 Plant competition experiment

its minerals before spring. This cheap method also maintains high humus content, good water retention and sound crumb structure of the soil.

Farms specialising in crop plants, and therefore lacking large supplies of animal manure, are often forced to use **artificial fertilisers** such as ammonium sulphate or sodium nitrate. This method is performed in spring so that the chemicals will not be leached out before the plants start to grow. Although such artificial chemicals are quick-acting, they are expensive and fail to add humus, so the soil still deteriorates.

Crop rotation
Different plants have different mineral requirements. Long-rooted plants remove nutrients from a deeper soil level than short-rooted plants. **Root nodules** on the roots of leguminous plants (e.g. clover, chapter 6) increase the nitrogen content of the soil. It is for these reasons that a 'four-course' system of **crop rotation** is so effective, for example:

wheat ⟶ turnips ⟶ barley ⟶ clover
1st year   2nd year   3rd year   4th year

In addition, parasites specific to a crop tend to die out during the three year absence of the host.

Plant competition

Crop plants (e.g. barley) often have to compete with vigorous **weeds** (e.g. poppy). When the barley plants shown in figure 35.4 are 'harvested' after several weeks' growth, the total dry mass for tray A is found to be greater than that for tray B. This is probably because the weeds have deprived the crop plant of one or more of many possible factors such as water, mineral salts, light and space to grow.

Pests

In a natural ecosystem, a balance exists between producers and consumers. If the green plants decrease in number then the number of dependent animals and parasites falls accordingly, allowing the plants to increase in number again and so on. However in a **monoculture** (a vast cultivated population of one crop plant), ideal conditions are presented to pests and parasites to feed and reproduce for several generations without running out of food.

Chemicals

In an attempt to control such populations of unwanted pests and weeds, chemical pesticides are used such as **herbicides** (which kill weeds), **fungicides** (which kill fungi) and **insecticides** (which kill insects).

Herbicides (selective weedkillers) are especially effective on broad-leaved plants (with vulnerable high-up growing points) as shown in figure 35.5 but have little effect on narrow-leaved grass plants (with ground-level growing points).

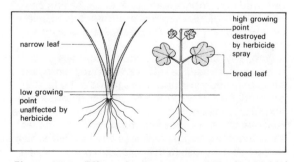

Figure 35.5 Effect of selective weedkiller (herbicide)

Many insecticides and fungicides are very effective at killing pests, but it has now been discovered that some of these chemicals can lead to some very serious **side effects**.

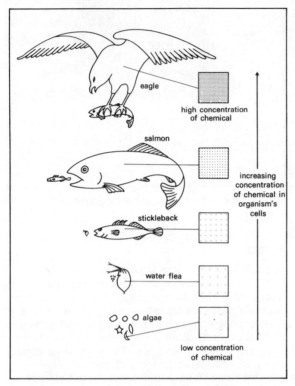

Figure 35.6 Accumulation of chemical in food chain

Effect of chemicals on food chains

Consider the food chain in figure 35.6. The producer, (unicellular alga) in a river, becomes contaminated with a very **low concentration** of chemical pesticide washed off neighbouring farmland by rainwater. The concentration increases however as many of the plants are eaten by the primary consumer (water flea) and built into its cells. Progression on up the pyramid of numbers leads to ever **increasing concentrations** of the chemical accumulating in living cells with finally the few large consumers at the very top being poisoned.

Similarly, although very small quantities of chemicals are used to kill insects on crops, predatory birds (e.g. sparrow hawks) are found to be fatally affected. Also, new strains of pests **resistant** to the chemicals have now been discovered. It is for these reasons of persistence and resistance that stable (**non-biodegradable**) pesticides such as **DDT** which neither decompose in the organism nor in the soil, have been banned in many countries.

Biological control

An example of this alternative method of pest control is releasing ladybirds on to a crop infested with greenfly. Once all the greenfly have been eaten, the ladybirds either die or fly off to another habitat. Thus the pest problem has been cured **naturally** and no chemical residues are left to pollute the ecosystem.

Revision questions

1 Starting with 'saprophytic bacterium', arrange the following bacteria in the correct order in which they would be met by an atom of nitrogen passing round the nitrogen cycle: *Nitrobacter, Rhizobium, Nitrosomonas*, saprophytic bacterium, denitrifying bacterium.
2 Briefly explain why:
   (a) manure is ploughed in during autumn, but fertiliser sprays are not applied until spring.
   (b) leguminous plants (e.g. clover) improve soil fertility.
   (c) some plants are severely affected by herbicides while others remain unharmed.

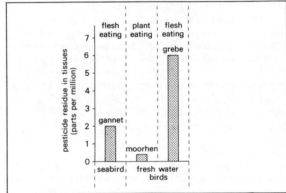

Figure 35.7 see question 3

3 Figure 35.7 shows the concentrations of pesticide residue found in the muscle tissues of certain water birds.
   (a) Account for the difference in concentration found between grebe and moorhen.
   (b) Suggest why gannets are less severely affected than grebes.
4 Mosquito larvae in rice paddy fields can be controlled by applying a film of oil onto the surface of the water. An alternative method of control involves the release of carp fish into the fields.
   (a) Which of these is a method of biological control?
   (b) Suggest how it works.
   (c) State 2 possible reasons why this method is preferable to chemical control.

36 Man and micro-organisms

Micro-organisms are tiny living things. They are often unicellular and can only be observed in detail under the microscope. The examples included in this chapter are **fungal moulds**, **bacteria** and **viruses**.

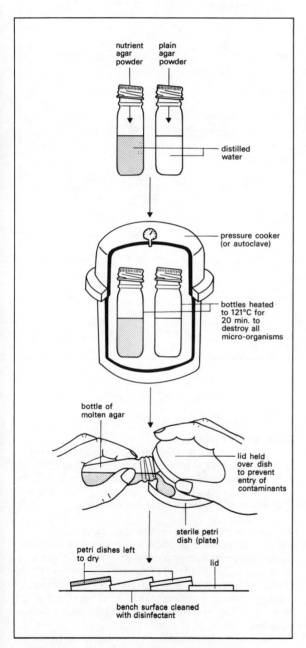

Figure 36.1   Preparation of sterile agar plates

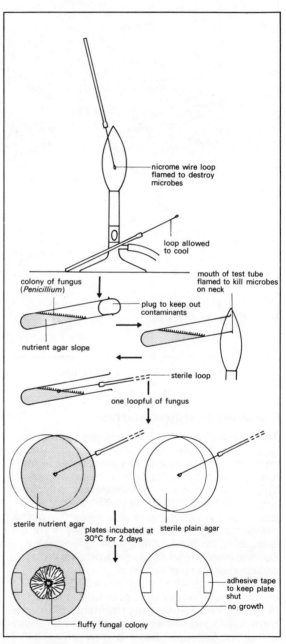

Figure 36.2   Growth of a fungal culture

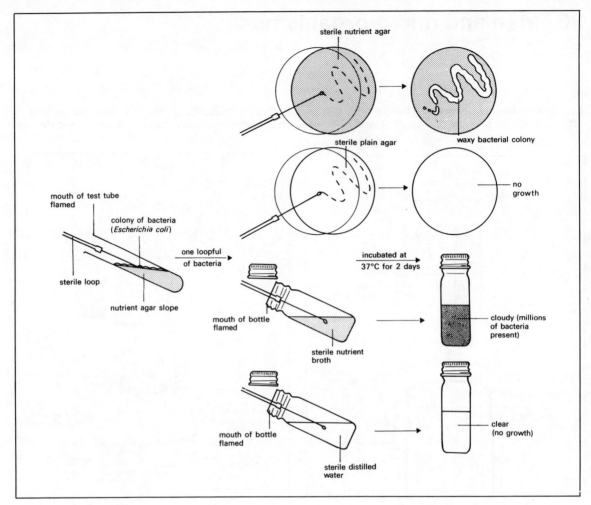

Figure 36.3 Growth of a bacterial culture

Culturing micro-organisms

The preparation of sterile agar plates is shown in figure 36.1. **Sterile techniques** are employed in an attempt to ensure that at the start of an experiment the agar is sterile (i.e. contains no micro-organisms). The agar is then inoculated with the particular micro-organism that is to be cultured. The further sterile techniques shown in figure 36.2 are carried out to prevent foreign micro-organisms entering and contaminating the plate while it is being inoculated. Despite these precautions **contaminants** do occasionally get into cultures.

From the experiments in figures 36.2 and 36.3 it can be concluded that growth of fungi and bacteria only occurs in the presence of **nutrients**. (These give the micro-organisms an energy source and a supply of building materials). Bacteria grow successfully on solid (agar) medium and in liquid (broth) medium.

Effect of pH on growth
When one sterile nutrient agar plate is set up at pH 3 (acidic), another at pH 7 (neutral) and a third at pH 10 (alkaline) and each inoculated with a micro-organism, most growth is found to occur at pH 7 and little or no growth at the other two pH extremes.

Effect of temperature on the souring of milk
From the experiment shown in figure 36.4 it can be concluded that bacteria present in fresh untreated milk are unable to thrive at extremes of temperature, but quickly multiply in warm milk, turning it sour. **Pasteurisation** (heating milk to 72°C for 15 seconds) kills most, but not all bacteria and effectively delays souring.

The effect of temperature on the growth of micro-organisms in general is summarised in figure 36.5.

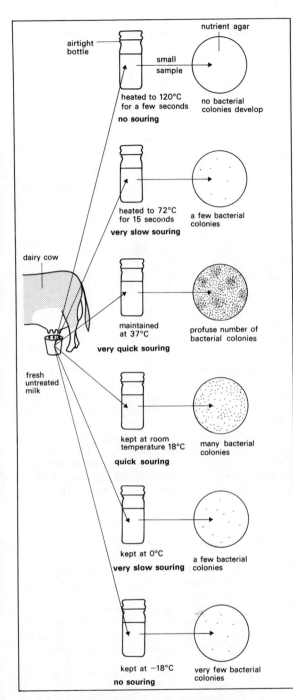

Figure 36.4 The souring of milk

Contamination of food by micro-organisms

Many micro-organisms produce air-borne **spores** which land on exposed foodstuffs and make them go bad. Techniques employed to prevent such food spoilage are summarised in table 36.1.

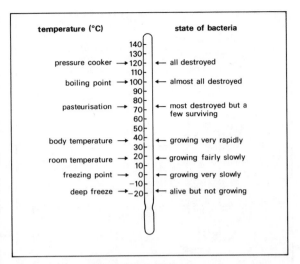

Figure 36.5 Effect of temperature on bacterial growth

technique	effect	example
high temperature	all micro-organisms killed	canned food
pasteurisation	most micro-organisms killed	milk, wine
low temperature storage	growth of micro-organisms prevented	dairy produce
drying	growth of micro-organisms prevented by lack of water	cornflakes
addition of sugar	growth of micro-organisms prevented by lack of available water	jam
addition of vinegar	growth of micro-organisms prevented by low pH	pickled onions

Table 36.1 Prevention of food spoilage

Natural occurrence of micro-organisms

The experiments in figures 36.6 and 36.7 show that human skin and fresh soil harbour many micro-organisms; roasted soil however is sterile. A similar experiment using fresh pond water shows that it also contains many bacteria. It is for this reason that drinking water is **chlorinated**. It can also be sterilised by **boiling**. The experiment in figure 36.8 shows that air also contains many micro-organisms; their growth is however inhibited by **disinfectant**.

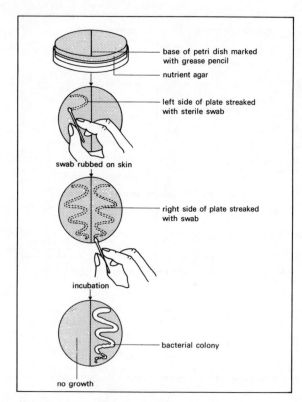

Figure 36.6   Bacteria from human skin

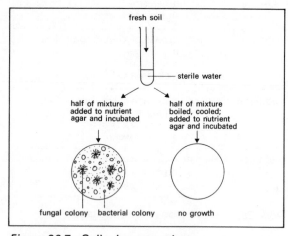

Figure 36.7   Soil micro-organisms

Clearly countless numbers of micro-organisms exist in our environment and many are present on the surfaces of our bodies. Fortunately the vast majority are harmless and easily removed by washing. A few however cause disease and are called **pathogens. Tuberculosis** is caused by a pathogenic bacterium, **athlete's foot** by a fungus and **poliomyelitis** by a virus. **Viruses** are disease-causing

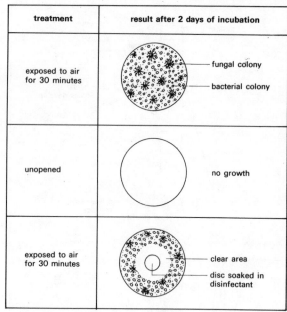

Figure 36.8   Micro-organisms from the air

agents which are much smaller than bacteria and are too tiny to be seen under a light microscope. They exhibit living and non-living characteristics and only reproduce within the living cells of other organisms.

When a pathogen enters the human body, the body's natural defence systems begin to operate. In particular this involves the white blood cells (see chapter 16). However sometimes these defences alone are unable to overcome the pathogen and the body requires the assistance of an **antibiotic**.

Antibiotics and the discovery of penicillin

In 1928 Alexander Fleming found a fungal contaminant growing on one of his plates of bacteria (figure 36.9). He noticed that the area around the fungal colony, instead of being cloudy with bacteria, was clear. He therefore concluded that some substance made by the fungus (*Penicillium*) was inhibiting the growth of the bacteria. This substance, an antibiotic, was later isolated and called **penicillin**. Tests showed that it was non-toxic to humans.

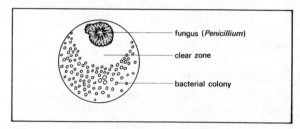

Figure 36.9   Fleming's famous plate

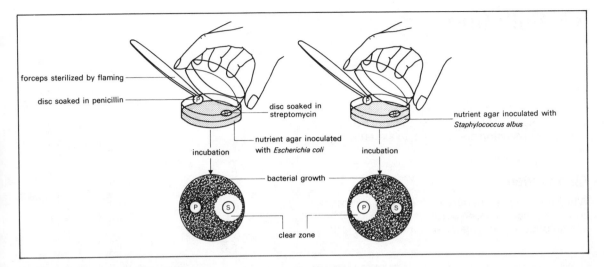

Figure 36.10   Action of antibiotics

Penicillin has now been used to cure many diseases such as pneumonia.

Several other antibiotics have been discovered. Each is a chemical produced by one micro-organism which prevents the growth, and may cause the death, of other micro-organisms. The experiment in figure 36.10 shows that each antibiotic is **specific** in that it only inhibits the growth of certain types of bacteria.

Useful applications of micro-organisms

Breadmaking
The fungus yeast added to uncooked dough gives off CO_2 bubbles during respiration. This makes the dough rise. It is then baked at a high temperature which kills the yeast.

Fermentation
In the absence of oxygen, yeast respires anaerobically (see chapter 11) producing alcohol and CO_2. Thus yeast converts malt solution into beer and fruit juice into wine.

Cheese-making
A particular strain of bacteria is added to milk which is made to curdle. The bacteria then act on the milk protein (casein) forming cheese.

Revision questions

1 Copy and complete the following table.

sterile technique	effect
swabbing work surface with disinfectant	
heating glassware in autoclave	
holding lid over open petri dish	
flaming wire loop	
flaming mouth of culture tube	
applying adhesive tape to plates	
washing hands before and after experiment	

2 State one way in which the appearance of a bacterial colony differs from that of a fungal mould.

3 Figure 36.11 shows a plate of nutrient agar smeared with fungus. Four different types of bacteria were smeared at lines 1–4 and the plate incubated for 48 hours. Account for the subsequent bacterial growth.

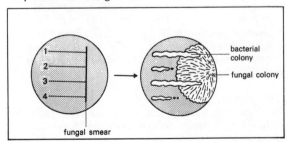

Figure 36.11   see question 3

37 Pollution

Environmental pollution is the contamination of our surroundings by substances which harm living things often causing disease and, in extreme cases, death.

Air pollution

When coal and oil are burned, they produce **smoke** containing soot (**carbon** particles) and gases (CO_2 and **sulphur dioxide**) which pollute the atmosphere.

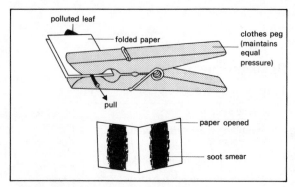

Figure 37.1   Air pollution of plants

Effect of soot on leaves

When the apparatus shown in figure 37.1 is used to compare privet leaves from a country area with those from an industrial area, the latter are found to bear much more soot. This outer layer of dirt reduces the amount of light reaching green cells and blocks stomata preventing CO_2 intake. Photosynthetic rate is therefore slower, growth is retarded and the plant remains stunted and small.

Sulphur dioxide

Even in low concentrations, this gas aggravates human **respiratory** ailments. It also causes leaf damage to plants and therefore reduces crop yield. **Lichens** (see chapter 6) are especially sensitive to sulphur dioxide and their almost total absence from many cities (see figure 37.2) indicates atmospheric pollution by this harmful gas.

Car exhaust fumes

Petrol contains **lead** additives. As a result, traces of poisonous air-borne lead are released from car exhausts.

Another product of petrol combusion is **carbon**

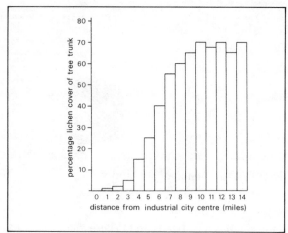

Figure 37.2   Effect of sulphur dioxide on lichen growth

monoxide. This colourless, odourless gas is poisonous since it combines with haemoglobin (see chapter 16) and impairs the blood's oxygen-carrying capacity.

Smog

In Britain, most smoke and fumes are quickly lost to the upper atmosphere. Air pollutants may, however, gather at ground level if a **temperature inversion** (figure 37.3) occurs. If fog (a cloud of water droplets) combines with this smoke, **smog** is produced. The combination of bright sunshine, industry and motor cars results in daytime smog often forming in Los Angeles, USA. 4000 Londoners died of respiratory diseases following the deadly smog of 1952.

Smokeless zones

These have been established as a result of the **Clean**

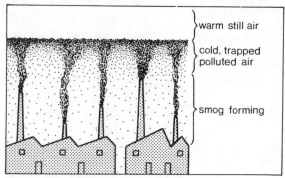

Figure 37.3   Temperature inversion

Water pollution

This occurs when **sewage**, **excess fertiliser**, **industrial waste** or **oil** is added to a river or to the sea.

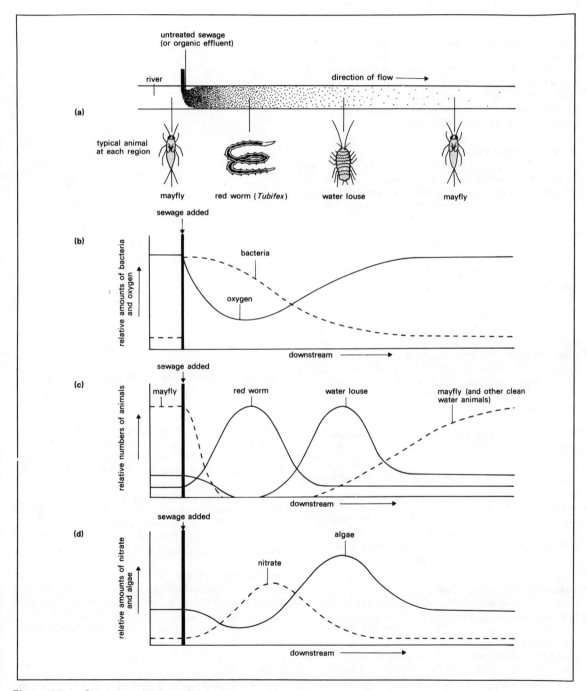

Figure 37.4 Organic pollution of a river

Sewage

Raw sewage consists of water containing **organic wastes** (faeces, soap, detergent, food fragments etc.). At a sewage works, this mixture is purified by micro-organisms (e.g. bacteria and *Paramecium*) which feed on it and break it down into harmless substances. When these products (the effluent) are passed into the local river, they do not affect the health of the river.

In some densely populated areas, however, the sewage works are overloaded. The **effluent** now passed into the river contains untreated sewage and many bacteria which feed on the suspended particles of sewage multiply rapidly. Soon the river's oxygen supply (figure 37.4b) is exhausted by this vast population of respiring micro-organisms.

Effect of oxygen shortage

Deprived of oxygen, clean water animals (e.g. **mayfly**, figure 37.4a) die. Tolerant species (e.g. **redworm**) however survive since they possess **haemoglobin** which enables them to make the best possible use of any tiny traces of oxygen present. Their population size normally increases (figure 37.4c) due to lack of competition in oxygen-deficient water. As the water begins to return to normal, semi-tolerant species (e.g. **water louse**) and then clean water species (e.g. **mayfly** and **fish**) reappear.

Effect of excess minerals and fertilisers

Sewage and detergents broken down by bacterial action release **nitrates** and **phosphates**. The high concentrations of these minerals that accumulate in a polluted river, increase the growth of algae (figure 37.4d). As a surface mat of algae forms, it prevents light reaching submerged plants which die and further deprive the river animals of oxygen.

An **algal bloom** also occurs when **excess fertiliser** (containing **nitrate**) is washed out of agricultural land into neighbouring rivers and lochs by rain.

Determination of oxygen content of polluted water

Ferrous sulphate (see figure 37.5) reacts with **oxygen** in water and only decolourises the dye when the oxygen supply is exhausted. It is found that less ferrous sulphate is required to decolourise the dye in polluted water than in tap water showing that polluted water contains less oxygen.

Indication of faecal pollution

Water polluted with human **faeces** always contains harmless bacteria called *Escherichia coli*. When special culture medium (McConkey's broth) is inoculated with *E. coli*, the broth changes from clear purple to cloudy yellow. If a water sample also brings about this colour change (figure 37.6), it must contain *E. coli* indicating that it is **polluted** with **faeces**. Although *E. coli* is harmless, faecally polluted water

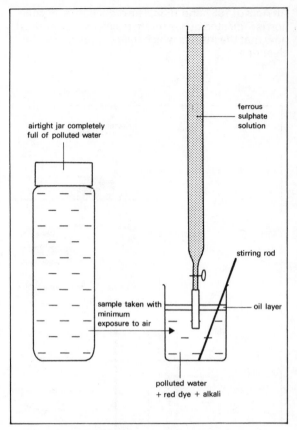

Figure 37.5 Oxygen content of polluted water

is unfit to drink because it may also contain harmful pathogenic micro-organisms.

Industrial wastes

When a factory discharges untreated **organic effluent** (e.g. paper fibres) into a river, bacterial growth is again promoted and the series of events already outlined is repeated. However the river will eventually recover.

When industrial works discharge poisonous **inorganic wastes** (acid, alkali, copper, lead, mercury etc.) into a river, no forms of life can survive, everything dies and the river becomes **biologically dead**.

Radioactive wastes

Certain radioactive wastes are especially dangerous since they continue to give out harmful radiation for many years. To prevent pollution these are often sealed in lead tanks and dumped into the deepest parts of the ocean.

Detergents

Excess **detergent** can reduce the efficiency of a sewage works by killing the useful micro-organisms.

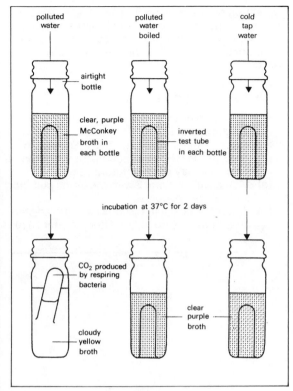

Figure 37.6   Indication of faecal pollution

If discharged into a river and whipped up into foam, detergent reduces oxygen uptake at the water surface depriving the river animals below.

Oil
Oil spillage from tankers at sea immediately affects those birds that spend much time on the water surface. The oily feathers lose their **insulatory** effect and the bird dies of **exposure** or becomes **poisoned** as it preens its clogged feathers with its beak. Oil washed ashore pollutes the animals and plants of the rocky shore. Methods used to clear oil slicks include adding **sand** to sink the oil and setting **fire** to it.

Noise pollution

This occurs when people are subjected to **noise** which is unpleasant because of its excessive volume, duration or shrillness. Every year thousands of industrial workers become partially or totally **deafened** by factory noise. For most other people, the major sources of unacceptable noise are motor traffic and aircraft.

Revision questions

1 Name 3 substances which cause atmospheric pollution.
2 Why do Japanese policemen often have to use oxygen masks while on traffic duty in the busy streets of central Tokyo?
3 State 2 sources of the minerals that lead to the formation of an algal bloom in a river.
4 In what way do **(a)** submerged plants and **(b)** river animals suffer as a direct result of an algal bloom?
5

water pollutant	possible method of control
excess detergent	use of minimum during washing
	decomposition by micro-organisms
	sinking, burning and prevention
	recycling and alternative means of disposal
	use of minimum on fields near rivers

Copy and complete the above table using the following examples:
inorganic chemicals; organic effluent; excess fertiliser; oil at sea.

38 Conservation

The size of a population of wild animals is normally kept stable by factors such as availability of food, incidence of disease and numbers of predators. In man, however, a **population boom** has occurred during the last century (see figure 38.1). Survival of so many people is due mainly to our exceptional brainpower! This has enabled people to improve public health dramatically and develop vast areas of land for food production.

Unfortunately the ever expanding world popula-

119

tion in turn makes ever increasing demands on the environment (with more and more land being cleared for farming) and on the world's supply of natural resources.

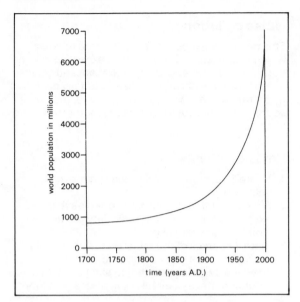

Figure 38.1   Increase in world population

Non-replaceable resources

Fossil fuels (**coal**, **oil** and **gas**) and essential elements (e.g. **lead**) are examples of such resources. It is vital that these finite resources are used carefully and not wasted, because when they become exhausted they will never be available again. Substances should be **re-used** wherever possible and waste materials containing useful materials should always be **re-cycled**. It is also essential that mankind continues to develop alternative sources of energy.

Replaceable resources

It is true that when trees are felled, more can be planted to replace them. However the consumption rate of timber must not be allowed to exceed the replacement rate. The world's **forests** are essential for the removal of CO_2 and the addition to the atmosphere of **oxygen** by photosynthesis. In Britain the **Forestry Commission** is responsible for the conservation and maintenance of our forests.

It is true that **aquatic animals** will breed again and replace those removed by man. However if the seas are overfished and whales are slaughtered indiscriminately, these animals will become extinct unless they are protected. Some rare animals and plants are already protected by law and others by living in special areas called **nature reserves**. However, conservation does not simply mean the

protection of rare species. It means the intelligent management and, where possible, improvement of our environment. It involves the controlled use of our natural resources. It is vital to man's survival on this planet.

Revision questions

1 Give 2 examples of conservation of the environment practised in Britain.
2 If the world population continues to grow at the present rate, it is thought that the trends graphed in figure 38.2 may occur in future years.
   (a) Match curves 1 and 2 with the terms pollution and natural resources.
   (b) Suggest 3 possible reasons for the predicted drop in world population after the year 2020.

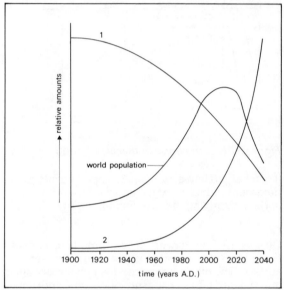

Figure 38.2   see question 2

3 Figure 38.3 shows the message on a recent popular badge. Why should man be thankful to trees?

Figure 38.3   see question 3

Index

egg
  animal 77, 78, 81
  plant 87
elbow 64
Elodea 37, 40, 41
  bubbler experiment 37
  cytoplasmic streaming 48
embryo 80, 83, 84
  plant 87, 89
  sac 87
endocrine gland 77
endoskeleton 60, 65
endosperm 87, 90
  nucleus 87
energy
  chemical 13, 35
  food 8, 9, 11
  food chain 42, 43
  heat 33, 34
  light 13, 35, 39
  release 33, 34
  requirements 12
enzyme 19, 20
  digestive 22–4
  lock-and-key 20
  photosynthesis 39
  respiratory 33
  -substrate complex 20, 38
Escherichia coli 112, 115, 118
ethanol 34, 36
excretion 51
excretory system 7, 52
exoskeleton 1, 2, 60, 65
expiration 30
extensor muscle 65, 74
eye 69, 70

faeces 24
  pollution 118
fat 9, 12
  digestion 24
fatty acid 9, 24, 25
fermentation 34
ferns 1
fertilisation 78, 81
  external 78
  internal 79, 87
fertiliser 108, 109
  pollution by excess 118
fibrin 50
fibrinogen 50
filter feeder 15, 16
fingerprint 94, 95
fishes 3
flexor muscle 65, 75
flower 1, 8, 86, 87
fluoride 15
foetus 83, 85
follicle stimulating hormone (FSH) 76, 77
food
  additives 10
  calorimeter 11
  organic 8, 13
  shortage 12
  tests 10

food chain 42, 43, 100, 101
  effect of pesticide 110
food web 42, 43
fructose 9
fruit 87
fungicide 109
fungus 1, 13, 111, 114

gall bladder 23, 24
gamete 78, 87
gaseous exchange 28, 31, 39
gastric juice 22, 23
gene 95–7
genetics 96
genotype 96, 97
genus 4, 5
geotropism 67
germination 89, 90
gestation 83
gill
  fish 31, 32
  mussel 15, 16
  tadpole 83
glomerulate filtrate 52, 53
glomerulus 51, 53
glucose 8, 9, 23
  conversion to glycogen 25
  energy 33
  oxidation 33
  photosynthesis 39
  reabsorption 52, 53
  transport 50
glucose-1-phosphate 38
glycerol 9, 24, 25
glycogen 8, 9, 25, 76
grafting 92, 93
greenfly 18, 100, 101, 110
guard cell 39, 55
gymnosperms 1

habitat 100, 107
haemoglobin 31, 49, 118
haemolysis 46
haploid 96
heart 7, 48
hepatic portal vein 24, 25, 49
hepatic vein 25, 49
herbicide 109
herbivore 14, 42
heterotroph 13
heterozygous 97
hip 64
homeostasis 77
homozygous 97
hormone
  animal 76, 77
  plant 67
  transport 50
housefly
  feeding 15
  reproduction 79
humus 101, 103–4, 108–9

ileum 23, 24, 51
indicator
  bicarbonate 28, 40, 41

congo red 47
universal 21, 29
indole acetic acid (IAA) 67, 68
insects 1, 2
  biting 15
  gaseous exchange 32
  sucking 15
insecticide 109
insectivorous plant 102
inspiration 30
insulation 9
insulin 76
intestine 7, 23, 24
invertebrate 60
  limb 65
iodine solution 5, 10, 19, 36, 38, 90
iron 10, 24, 49

joint 63
  ball and socket 64
  hinge 64
  immovable 63
  jaw 14
  peg and socket 65
J-tube 27, 28

key
  branch 1
  paired statements 4
kidney 7, 51, 52
  tubule 51, 52
kilojoule (kJ) 11
knee jerk 74, 75

lactic acid 34
ladybird 100, 101, 110
larva 82
lead 116, 118, 120
leaf 7, 8, 39, 55
  compound 8
  foliage 89, 90
  simple 8
  variegated 36
leguminous plant 17, 109
lens 69, 71
  accommodation 70
lenticel 32
lichen 17, 103, 116
life cycle 81, 97
ligament 64
  suspensory 69
lignin 59
lime 104, 106
lime water 27, 28, 34
limiting factor 37
Linnaeus 4
lipase 21, 22, 24
liver 7, 24–6, 49
loam 106
locust
  feeding 15
  metamorphosis 82
long sight 71
loop of Henle 53
lungs 7, 29, 30, 49
luteinising hormone (LH) 76, 77